SCIENCE
BY SIMULATION

Volume 1: A Mezze of Mathematical Models

SCIENCE
BY SIMULATION

Volume 1: A Mezze of Mathematical Models

ANDREW FRENCH
Winchester College, UK

 **World Scientific**

NEW JERSEY · LONDON · SINGAPORE · BEIJING · SHANGHAI · HONG KONG · TAIPEI · CHENNAI · TOKYO

Published by

World Scientific Publishing Europe Ltd.

57 Shelton Street, Covent Garden, London WC2H 9HE

Head office: 5 Toh Tuck Link, Singapore 596224

USA office: 27 Warren Street, Suite 401-402, Hackensack, NJ 07601

Library of Congress Cataloging-in-Publication Data

Names: French, Andrew, author.

Title: Science by simulation / Andrew French, Winchester College, UK.

Description: New Jersey : World Scientific, [2022] | Includes bibliographical references and index. |
    Contents: volume 1. A mezze of mathematical models.

Identifiers: LCCN 2021041907 | ISBN 9781800611078 (hardcover) | ISBN 9781800611214 (paperback) |
    ISBN 9781800611085 (ebook) | ISBN 9781800611092 (ebook other)

Subjects: LCSH: Simulation methods.

Classification: LCC T57.62 .F74 2022 | DDC 658.4/0352--dc23/eng/20211004

LC record available at https://lccn.loc.gov/2021041907

British Library Cataloguing-in-Publication Data

A catalogue record for this book is available from the British Library.

For any available supplementary material, please visit
https://www.worldscientific.com/worldscibooks/10.1142/Q0327#t=suppl

Desk Editors: Balamurugan Rajendran/Michael Beale/Shi Ying Koe

Typeset by Stallion Press
Email: enquiries@stallionpress.com

To my family, to wise, beautiful and ever tolerant Lucinda, to my friends Matt, Lynne, Walter, Veerle, Greg, and all my students, teachers and colleagues from Carisbrooke, Cambridge, Cowes, Chelmsford, Dorchester, Sherborne and Winchester.

Foreword

Science progresses by seismic shifts in perspective: Thomas Kuhn argued that science does not evolve steadily towards truth but rather by paradigm shifts when current theories cannot explain some phenomenon, and a new theory and viewpoint is proposed. In addition to the concepts and ideas of science, how scientists perform science has undergone analogous upheavals over time. Arguably, the first paradigm shift, from natural philosophy to science, is characterised by the transition from purely empirical evidence gathering to scientific theory building. In contrast, the next two scientific method paradigm shifts, of computational science and data-intensive discovery, have been marked by an unprecedented (and increasing) rate of progress, driven by the exponential Moore's Law growth of computing hardware over the past half-century. However, this pace and approach has not been mirrored in education. Students are typically taught techniques for tackling STEM problems using the pre-computing zeitgeist of the early 20th century. Traditional courses at secondary and early university levels may ill-prepare students to solve current problems, as well as limit the chance to learn and understand, by quick simulation, well-established scientific models.

Andrew French's new book and the accompanying materials on his *Eclecticon* website are now where I would recommend all young scientists begin their journey to learn how to do modern science. *Science by Simulation* introduces readers to the modelling process as an integral part of the scientific method and then showcases a wide-ranging selection of examples from across all the STEM disciplines, demonstrating the universality of the approach. All that a keen Sixth form or undergraduate student requires is access to a computer and he/she can construct and manipulate their own version of the recipes in the text. The only true way to learn is by doing and experimenting — something that Andrew encourages. Indeed, Andrew and the British Physics Olympiad have successfully trialled some of the material in a hugely popular online course this year, culminating in a new computational competition, which is the first in the world for Sixth form students.

I encourage the reader to join in, switch on their computers, and take a step into the new computational paradigm of science.

Dr Anson Cheung
British Physics Olympiad
April 2022

Preface

A Mezze of Mathematical Models is Volume 1 of *Science by Simulation*. It is a recipe book of mathematical models that can be enlivened by the transmutation of equations into computer code. In this volume, the examples chosen are an eclectic mix of systems and stories rooted in common experience, rather than those normally associated with Physics, Chemistry or Biology in isolation. This approach will hopefully provide a tangible and readily accessible context for the development of a wide range of interconnected mathematical ideas and computing methods that underpin the practice of Science.

Particular thanks to Dr Anson Cheung and Robin Hughes from the British Physics Olympiad, and Jeremy Douglas and Tony Ayres from Winchester for so many inspirational conversations. In assisting my first tentative steps in the publishing process I would like to thank Dr Forrest Sheldon from the London Institute for Mathematical Sciences and Michael, Bala, Shi Ying and Laurent from World Scientific. Most importantly I would like to offer my gratitude to *sensei* John Cullerne for the amazing opportunity to collaborate on the *Epidemiology of Eyam*, and especially Chas McCaw for encouraging me to transform a very speculative idea at a summer track & field match into the content of this book.

Science by Simulation is dedicated to the next generation, and particularly my nieces and nephews Dominic, Sophie, Jamie, Charlie, Esme and Sebastian.

I am very grateful to the Governing Body and Headmaster of Winchester College for granting a Sabbatical in 2021 to complete the final phase of *A Mezze of Mathematical Models*.

About the Author

 **Dr Andrew French** has taught Physics and Mathematics at Winchester College since 2011. Previously, he taught at Sherborne School, and worked in industry for eight years as a Radar Systems Engineer on a wide range of projects: from meteorological sensors, to wind farms, to signal processing algorithms associated with marine and land-based systems. While in industry, he completed a PhD from University College London on aspects of phased-array radars. He studied at Christ's College, Cambridge, from 1997–2002, culminating in a Master's in Experimental & Theoretical Physics. He also holds a postgraduate Master of Philosophy in Fluid Dynamics from the BP Institute in Cambridge and a PGCE in Secondary Mathematics teaching from Southampton University. Dr French is currently working on a number of educational outreach and research projects: *The Eclecticon* resource website, *Science by Simulation* series of books, and he has also co-authored several recent papers in *Physics Education* relating to epidemiology and numerical methods in introductory calculus teaching.

Contents

Chapter 1

Introduction

1.1 Science, Models and Maths

The Oxford Dictionary of English defines *Science* as "the intellectual and practical activity encompassing the systematic study of the structure and behaviour of the physical and natural world through observation and experiment." This description is a beautifully compact *what*, but not so illuminating when in comes to fleshing out the *how* of Science. Assigning a technical name to a physical phenomenon or property (e.g. 'Energy', 'Force', 'Light', 'Momentum', 'Coronavirus', etc.) may be a sensible starting point; categorizing and simplifying the near infinite complexity of human experience into structures of similarity that permit systematic study. However, the practice of Science goes much further. A scientist is a builder of mathematical models which comprise rigorously defined quantitative (i.e. numerical) relationships between physical quantities that can be *measured*. Physical laws are expressed using the universal language of Mathematics. This removes ambiguity or bias resulting from the cultural nuances of spoken languages, and enables a clarity of discourse between scientists. Physical laws are formed into *equations*, which can be rearranged. This means a difficult to measure quantity (e.g. magnetic field strength) can be calculated from an easier to measure observable (e.g. the voltage produced from a Hall Probe semiconductor device). Many equations of Science are expressed in terms of a *rate of change of a quantity with time*. For example, the gradient of a velocity vs time graph is proportional to the force applied to an object, if its mass is constant. If one can sum infinitesimal changes of a quantity over time, from a known starting point (this process is called *integral calculus*), then knowledge of the gradient of the time variation of a quantity can enable direct calculation of the quantity itself. This means, subject to the errors associated with our measuring process and their growth, we can predict the future and know the past.

1.1.1 *Forces to be reckoned*

Knowledge of the gravitational forces upon the oceans of the Earth due to the Sun and the Moon enables us to predict tidal variations with exquisite precision, and from the same models we can reliably send probes to Pluto, land on comets and plan manned missions to Mars. Knowledge of electrical and magnetic forces allow us to design motors and generators that enable lighting, heating, communication, construction and efficient movement over large distances. This mastery of electromagnetism, combined with an understanding of the Quantum Mechanics and atomic physics of the smallest of structures, enables humanity to construct microscopic electronic circuits, and therefore macroscopic (i.e. person-sized) computing machines of awesome intricacy and calculating power. An ability to explore and control the microscopic and nanoscale[1] yields power over cells, bacteria and viruses, and therefore enables a potential freedom from disease and other biological miseries. True artificial intelligence in the form of conscious robots may still be in the realm of science fiction, but augmentation of the human mind by technology has been widespread since the invention of mechanical calculator and printing press, and entirely ubiquitous in the modern era of cheap Terabyte hard drives, Gigahertz processors and wireless broadband internet. I could fill a smartphone with an entire library of knowledge, without changing its physical mass or volume. Our ancestors, even a hundred years ago, would be awestruck by this miracle. Two or three hundred years ago they would probably classify it as magic and accuse you of possessing demonic powers. At the smallest of scales, an understanding of nuclear forces and particle physics yields insight into the genesis and possible lifecycle of the entire Universe, and the root structure of all matter. It also reveals the primary source of energy, which in the case of fusion (which powers the Sun and all other stars) may some-day replace our dependence on fossil fuels. Not only do fossil fuels pollute our ecosystem, they offer a million times less energy per unit mass than nuclear processes.[2]

1.1.2 *A wider world of examples*

In reading this book you will have already experienced a dazzling myriad of human inventions that only work because they have been designed to exploit physical laws. Your reading location is hopefully warm and dry. You have a lighting system that

---

[1] A nanometre (nm), about 10 times the radius of a hydrogen atom, is one *billionth* of a metre. (10^{-9}m).

[2] If the annual energy demand in the UK is about 10^{19} J per year, this equates to just 30 tonnes of deuterium and tritium fuel needed to power a nuclear fusion reactor. The reaction $^{2}_{1}\text{H}+^{3}_{1}\text{H}\rightarrow^{4}_{2}\text{He}+^{1}_{0}\text{n}$ releases $\Delta E = 2.82\times10^{-12}$ J. This might not sound like a lot, but per unit mass of inputs, it is 3.37×10^{14} Jkg^{-1}. By comparison, the combustion of coal releases about 3.5×10^{7} Jkg^{-1} and natural gas about 4.9×10^{7} Jkg^{-1}.

turns on almost immediately and, if you use a modern LED bulb, produces very little waste heat. You wear clothes that are machine woven to microscopic smoothness, are breathable, light, and didn't cost you a prior encounter with a massive, hairy and murderously angry beast. If you get sick or injured, chances are you can take appropriate drugs to cure your ailment, and use appropriate cleaning and sealing systems to help wounds heal without excessive scarring or infection. You may be sipping a coffee whose beans were harvested half way around the world, and via a complex system of electronic, paper and metal-based monetary exchanges, is available at very low cost. Everywhere you look there are vast riches yielded by application of scientific principles, and therefore an immediate diversity of context for tangible examples of how Science works. The key ideas of *Science by Simulation* are as follows.

1.1.3 *Learn to build models*

To fully appreciate Science *you have to build the associated mathematical models yourself.* The assembly of the private universe of a mathematical model is an absorbing and highly creative task. It is also practical and may involve the use of computer technology to perform repeated calculations, plot graphs, make spreadsheets, write code and construct graphical user interfaces, etc. Once a model has been created, one can investigate if it correlates with the real world via a suitable experiment. A successful prediction of measurable phenomena from prior calculation is a fantastic thrill. You are clairvoyant, oracle and prophet for the day. But your revelations are grounded in reality. If a model does not correlate, we rethink the model rather than assume reality is temporarily misbehaving.

1.1.4 *Master mathematics*

The mathematical details are important. You can't gloss over them if you plan to build a computer model to calculate the elevation angle of the green arc of a rainbow, or the predicted number of plague infectives in a month, or the shortest distance between London and Hong Kong. The real power of a computer is its ability to perform rapid calculations, billions a second, without getting bored and making errors. However, a computer needs to be programmed what to do, step-by-step. With some caveats regarding the ongoing development of artificial intelligence, in most cases only a human programmer will do. The intricate details of computational recipes, the typography of their representation, and the logic they contain, can be exceedingly beautiful. Algorithms are poems, but with immediate and unambiguous practical use. Unfortunately there appears to be a cultural rule in pre-University education that in order to open up Science to a wider audience, there must be a necessary diminution of mathematical content because of the perceived difficulty. Alas this perhaps says more about the personal anxieties of the educators than the potential of the educated. Calculus, an essential feature of the description of most physical laws, used to be taught at O-level (14–16 years) in the UK, now it barely features in Advanced-level (16–18 years) pre-University courses. This is folly for many reasons, and it means students may begin a University course with a paucity of skills and a lack of rich experiences. I believe in human progress, and it seems bizarre that a grade A student of 2020 should be less capable than a grade A student of 1970. This is not true in music education, (good luck attempting a Grade 8 without being able to read music) so why do we allow such a dumbing-down in our educational core? This glossing-over-the-quantitative-details mantra may also cause unintended negative knock-on effects. Computer programmers that are content, perhaps by ignorance, to create vast edifices built upon black-box functionality, will produce even more bloated systems that are poorly understood when they don't work. And given our world is increasingly based upon information technology, this is a serious problem for future resilience and personal autonomy. Conversely, those that do know how to build systems from the ground up will command high economic value. An appropriate and timely national response to a natural disaster, a recession or a pandemic needs to be based upon a rigorous and rational understanding. This cannot be achieved if anything mathematical brings you out in a cold sweat. And on a more positive note, in my experience of teaching, students really enjoy the process of model building. If as a teacher you explode the myth that maths can only be comprehended my macho otherworldly geniuses, you will bring great joy to your students as they assemble a bombproof step-by-logical-stepwise comprehension of the world. *The details matter, and these details are encoded in mathematics.* If a class struggles with mathematical ideas, choose a context which is more interesting and immediately accessible to them. Learners need an incentive which is more positive and powerful than a fear of failing exams and looking foolish relative to their peers. Curiosity and a desire to solve technical problems will do nicely. There is plenty of motivation for an understanding of quadratic equations, trigonometry and calculus if the successful running of a really interesting computer model demands it.

1.1.5 *Learn to code dynamic computer simulations*

The building of a workable model (i.e. with inputs, calculations and outputs) is the key intermediate step between the real world and the world of the laboratory experiment. Building the model provides the incentive for the acquisition of mathematical and scientific knowledge, and enables one to make sense of experimental work. Dynamic computer simulations[3] are a key educational element in modern science teaching, and are particularly useful for topics such as electricity that cannot be readily experienced in a direct physical way (as compared to mechanics, thermodynamics or optics). *Science by Simulation* is essentially a recipe book of computer

---

[3] A particularly fantastic set of simulations, many of which can be run in a standard web browser, have been developed by the PhET project of the University of Colorado. https://phet.colorado.edu/.

simulations. However, despite significant investment and development, and the gradual integration of dynamic simulation and datalogging technologies, there is often a significant separation between *creative* Computing and Science education. They are still often taught separately. This is as ridiculous as distinguishing geometry from algebra and trigonometry, or indeed 'Mathematical Physics' from Physics, as I have spotted in a few University course prospectuses. I'm not really sure what Physics without Mathematics can be, beyond a labelling of phenomena, and being awarded what amounts to an illusion of competence cannot be good for the future skills development of the next generation. I'm sure reputable Physics courses without the named Mathematics add-on actually have plenty of it, but the fact that the semantic separation exists should perhaps be considered as an ill omen, or indeed a 'frictionless inclined plane'... If one regards computing as calculation, use of information technology in a school or university laboratory is merely an extension of traditional on-the-board arithmetic. Learning to make computer models, based upon variable input parameters, also incentivises students to solve problems more generally. It therefore makes them better general-purpose problem solvers. This means a more rigorous and active use of *algebra*, since what is algebra other than a description of generic relationships between numbers, and therefore the syntactic basis of mathematical modelling.

1.1.6 *The power of context*

Do not underestimate the pedagogical power of context. If you choose a physical system wisely, you can use it as a vehicle to teach pretty much all of pre-University mathematics. The key criteria is that the system must be interesting, relevant and accessible. The first (and largest) chapter of this book *The Epidemiology of Eyam* is hopefully a reasonable example. You can explain the key epidemiological idea of Infectives, Susceptible, Dead and Recovered population subsets (and their characteristic curves) in a matter of minutes. However, to fully understand and characterize these curves is a journey through pretty much the entire syllabus. To construct a computer model of these curves is to learn the vast majority of spreadsheet and coding skills necessary for modern scientific work. In a sense, Science is *fractal*. You can choose pretty much any subset of the physical world, and within it contains all of Science. So choose wisely and aim for maximum enthusiasm for both teacher and student, but in essence the choice doesn't matter.

1.2 The Style and Structure of *Science by Simulation* (Volume 1)

This is the first volume of *Science by Simulation*. As the title *A Mezze of Mathematical Models* suggests, it is a deliberate mixture of contextualized examples of systems that can be modelled using mathematics, and simulated using computers. Whereas future volumes will focus upon the classic models of Physics (that every pre-University student ought to be intimately familiar with), this initial volume aims to develop a model-building style using a much wider range of examples, most of which are well outside the normal range of standard topics. A synopsis of each chapter is given below. We shall conclude this introductory chapter with two short examples (*Election Cups* and *Snails of Pursuit*) which hopefully illustrate the process of mathematical model construction and evaluation/visualisation using spreadsheets and computer programs.

1.2.1 *Chapter synopsis for A Mezze of Mathematical Models*

(1) **Introduction: Election Cups and the Snails of Pursuit.** Two short examples of (i) interesting and tangible context; (ii) mathematical model building; (iii) analysis and model visualisation using spreadsheets and computer programming.

(2) **The Epidemiology of Eyam.** Motivated by the tragic (and heroic) story of the impact of the Plague virus in the Derbyshire village of Eyam in 1666, a development of a set of differential equations which relate the Susceptible, Infective, Dead (+ Recovered) populations of this closed system. The use of William Mompesson's parish records enable modern data analysis methods to be utilised, and important parameters of the *Eyam equations* to be inferred. The Eyam model is then applied to the 2014 Liberia Ebola outbreak, and to the 2020 Coronavirus pandemic. The Eyam Equations are solved via numeric, and then 'semi-analytic' methods, and the solution mechanism is encoded in a computer application 'app' with a graphical user interface. In addition to continuous *deterministic* models, a discrete *stochastic* Eyam model is also developed and analysed. i.e. an integer number of persons with disease, (as opposed to an unconstrained decimal quantity) and the use of *random chance* in the determination of the spread of disease in any given time step. The latter results in a slightly different infective population vs time curve for each simulation run.

(3) **Holmes and Watson Meet Bayes.** The key ideas of *conditional probability* and *tree diagrams* are introduced, and yield mathematical tools for determining the probability of a particular hypothesis being true, *given* a set of observations. For example: the chance of having a disease, given the passing of a test that is quoted as being accurate to a certain percentage, which is typically less than 100%. In most situations this test-accuracy-percentage is the probability of passing a test, *given* samples that actually have the disease. *Bayes' Theorem* enables us to calculate what we actually want: the *likelihood of disease given a test pass*. It is assumed we already know the probability of a test pass given the presence of the disease, and indeed the probability of the disease in a patient expressing similar symptoms. These are the inputs to the Bayesian model. The possibly significant asymmetry between the outputs and the inputs of the

Bayesian model are very important in the general study of *inference*, i.e. how we know what we know given bounded uncertainties. It could therefore be argued that the Bayesian system of logic underpins the whole of Science.

(4) **May's Chaotic Bunnies.** Robert May's iterative model of how a population of rabbits might develop year-on-year is simple, yet has many very surprising features. By altering a single growth parameter, the population dynamics moves from extinction, to stable asymptotic populations, to cycles of fixed populations, and ultimately, via a cascade of bifurcating cycles, to essentially random behaviour. May's model, known as a *Logistic Map*, is a classic gateway to the key ideas of *Chaos Theory*, nonlinear dynamics and fractals.

(5) **Pendulums, Poincaré Diagrams and Strange Attractors.** This is the second of three chapters relating to aspects of Chaos Theory. This chapter develops a simple model of a swinging pendulum and demonstrates how even quite basic systems (two pendulums coupled together) can result in extraordinarily complex dynamics. Conversely, the *Strange Attractors* of Lorenz and Rössler illustrate how a certain 'geometric order' can manifest in systems which otherwise exhibit chaotic variations with time.

(6) **A Standard Atmosphere.** The variation of temperature and pressure in our multi-layered atmosphere is a readily accessible context for the study of key thermodynamic and ideal-gas concepts. If you have climbed a large mountain you will notice that pressure and temperature reduce significantly with altitude, as does the boiling point of water. Humidity will also have an effect, reducing the *lapse rate* of temperature as you ascend. To understand the atmosphere is to gain the basics of meteorology, vital for the science of weather forecasting, aeronautics, high altitude mountaineering, and skydiving!

(7) **The Subtlety of Rainbows.** Optics is one of the oldest scientific disciplines, and one which requires an application of both algebra and geometry in fairly equal measures. It it utterly embedded in daily experience, yet is sadly often neglected in many Physics courses. Descartes' beautiful model of a rainbow (both primary and secondary) is developed, which also involves an empirical analysis of the frequency dependence of the refractive index of water. The ray model of a rainbow is a very visual example of a macroscopic phenomenon (i.e. the coloured arcs observed in humid air when the sun is behind you) resulting from the microscopic phenomenon of light rays being internally reflected inside raindrops.

(8) **Exploring Julia's Fractals.** Infinitely complex and organic-looking fractal virtual worlds, resulting from a deterministic iterative transformation of *complex numbers*, is presented by means of a gallery of examples such as *The Mandelbrot Variations*. Armed only with a very mediocre computer and some basic coding skills, you can create these beautiful images (and their infinite variations) for yourself.

(9) **Radar, Chirps and Phased Arrays.** Humans, and most animals and insects, communicate wirelessly via sound and/or electromagnetic waves. This chapter describes the essentials of a Radar system, the benefits of frequency coded waveforms ('chirps') and how these signals might be processed. It also describes how an antenna comprising multiple radiation sources can be configured to steer a beam much faster than the antenna can be physically rotated.

(10) **Navigating the Sphere.** Navigation on a spherical planet requires the ability to: (i) realise (and demonstrate) that it is indeed spherical; (ii) calculate the radius of the planet; (iii) be able to plot circles on a spherical surface model, specify lines of latitude and longitude, 'great circles' and hence calculate the shortest distance between two points. In this chapter, we will look at the experimental methods of Eratosthenes and Al-Biruni for measuring the radius of the Earth, and how this relates to the Great Trigonometrical Survey of India in the nineteenth century. We will then develop a general method for specifying generic circles of all possible sizes and positions on a sphere, and apply this to classic Earth-based navigation problems.

(11) **Modelling Money and Mortgages.** Excessive debt, recession, rising house prices and a volatile stock market are all important personal and national problems in most modern societies. Ignorance of basic financial concepts will not help improve this situation. In this chapter we will develop realistic mathematical models of savings schemes and fixed-rate mortgages, and then virtually dabble in FTSE 100 stock trading by means of a *zbroker* algorithm based upon very simple statistical ideas.

(12) **Power to the People.** Between 2010 and 2019 the United Kingdom experienced *four* General Elections, and significant ideological turbulence. In this final chapter we will focus upon data analysis and information presentation to help make sense of the changing political landscape. We will also critique the democratic value of the first-past-the-post (FPTP) system adopted in the UK, and also, by means of a 'what if?' computer simulation, critique the possible impact of a proportional representation (PR) model.

1.2.2 *This doesn't look like a science book*

This is not a textbook. It is deliberately not tied to any particular syllabus, or indeed any higher educational course. In the real world of applied science, most graduates don't actually work in pure fields of Physics, Chemistry or Biology, but instead contribute to messy hybrids which also include human behaviour, money and design engineering. One of the main reasons for studying Science is that the Scientific Method of logical reasoning, constrained by experimental reality checks, is our only reliable schema for managing human affairs that offers the chance of long term individual human empowerment and improved life chances. If this statement sits uncomfortably with you, imagine your youngest child accidentally discovers a one-way time portal in her bedroom and finds herself stuck in Eyam in 1666, and without any antibiotics.

Applied rigorously, the Scientific Method can overcome groupthink and autocracy, emotive and blind faith driven illogicality, and a myriad of other cognitive biases that affect us all. Despite what an unmentionable UK politician recently quipped, the world needs more experts, particularly because experts (assuming they can keep their egos in check) are far more aware of what they *don't know* than a more blissfully ignorant layperson. As Richard Feynman almost certainly would have said, *Science is the calculus of doubt.* It is how we quantify uncertainty and enable us to make optimal decisions based upon incomplete information. A reliance on Science is also to be confident that we have acted rationally at a given time, even if our actions may prove to be less than desirable with the power of hindsight, and much more acquired knowledge. Without such confidence we are either cavalier, or paralysed into inaction. This book is about building mathematically rigorous models of a wide variety of (hopefully interesting) systems, and in doing do, the model builder becomes acutely aware of what the model may predict, to what level of certainty, and what assumptions are inherent in the model. Here are two examples:

In the *Eyam equations*, we assume a fixed population of *Infectives, Susceptibles, Dead* and *Recovered*. This may not be true at all in a country with a porous border. However, the general shape and structure of the predicted Infectives vs time curves will give more insight into large scale epidemiological trends then if we dispense with any modelling altogether and react purely emotionally to day-by-day stimuli and stresses. Without a mechanism to act strategically in the long-term, we are stuck in fire-fighting mode and will be even more open to the negative impact of future problems.

In Descartes' model of a rainbow we assume rain droplets are perfectly spherical. You might reasonably suggest that the polar nature of water molecules and the electrical nature of thundery weather may imply that this is not a particularly good assumption. Should we abandon the model because it is not perfect? Well, rainbows are commonly visible, and the model predicts the elevation angle of coloured arcs above the 'anti-solar-direction' to high precision. The beauty of Science over other systems of thought is that we can, via an experiment or direct observation, 'ask' reality to arbitrate whether our reasoning is likely correct, or flawed.

You are not going to be able to solve all the world's problems by reading this book, and it probably won't help you pass any exams. However, you might understand how rainbows work, what Chaos is and how to generate it, create beautiful and beguiling fractals, learn about navigation and the atmosphere, and be able to make more sense of virus-related national briefings or the state of your financial investments. In our increasingly ideologically digital world, it might even give you an edge when you next have to sensitively arbitrate between warring family members who have chosen to align with one of the annoyingly pervasive populist binary tribes such as

'Brexiteer' or 'Remainer' in the UK, or 'Tea Party Republican' and 'Green New Deal Democrat' in the USA, or perhaps even 'Boomer' vs 'Millenial' on both sides of the Atlantic. Wield the sword of data-driven logic to slay the puffed-up magic dragon of misinformation! If you read this book actively you will certainly become much more adept at model building and problem solving. And if you build all the models yourself, you will certainly develop expertise in scientific computing. For younger readers, I can't imagine well-honed skills in both of these aspects will harm your future career prospects, particularly in a world of artificial intelligence and broad spectrum embedding of information technology. In such a world, I'd want to be the programmer, not just the consumer, or worse, the programmed.

1.3 Election Cups

1.3.1 *Context*

I teach Physics at Winchester College, a venerable educational institution in the south of the United Kingdom that has been educating pupils since its foundation by William of Wykeham in 1382. The scholars of *College*, the oldest subset of the school, take a tough examination called *Election* prior to entry. One of my favourite pieces of the many historical artifacts acquired over the centuries is an ornate silver-gilt chalice know as the *Election Cup*. It was presented to Winchester in 1555 by Bishop John White of Lincoln[4] and is illustrated in Fig. 1.2(a). As a historical homage, I designed a recent Election Physics practical[5] to be loosely based upon the Election Cup, although my concept perhaps owes more to the children's song "there's a hole in my bucket, dear Eliza...". The idea is to investigate how the time taken for a cylindrical cup to sink, when placed in a large beaker of water, varies with the diameter of a circular hole drilled in the bottom of the cup. The experiment is simple and fun to perform, and although some manual dexterity is required, it requires very basic laboratory equipment. (Stopwatch, calculator, large water-filled beaker, mark-one eyeball). All bar one of the eight cups have the hole diameter printed on them. The idea is that students are guided through a process of careful experimental work and mathematical model building, which ultimately enables them to predict the missing hole diameter from the time taken for the unlabelled cup to sink. An interesting extension is to apply a small amount of liquid detergent to the water surface, which reduces the sinking time by a measurable amount.

The actual *Election* practical is aimed at precocious 12–13 year-olds, and the mathematical details are suitably calibrated. However, if we explore the entire process, this reveals many key principles of Physics: pressure in fluids, Archimedes'

---

[4] *Election Cup.* From *50 Treasures from Winchester College* [10, pp. 64–65].

[5] *Election Cups* was intended to be used in April 2020, but sadly could not proceed due to COVID-19. The practical has now been redesignated as a specimen paper for the practical science component of Election, and should be available from https://www.winchestercollege.org/.

principle of buoyancy, rate of change of momentum and the relationship to force, continuity of fluid, and the idea of a *vena contracta*, the natural radial contraction of fluid flow in the vicinity of a hole or *orifice*. To work out a relationship between sinking time and hole diameter requires basic calculus, and the final result turns out to be an inverse-square law, just like gravitation, electrical fields, radiative flux, etc. As illustrated in Fig. 1.3, the implementation of the model in a spreadsheet is a natural next step, and allows sinking time predictions to be made, and compared to measurements. The assessment of model to measurement correlation is via a line-of-best-fit to a suitably linearised version of the relationship between sinking time and hole diameter.

In summary, *Election Cups* is aimed as an example in miniature of the key analysis processes employed in pretty much all science experiments.

1.3.2 *Election Cups model of sinking time vs hole diameter*

The pressure difference ΔP between the bottom and the top of the hole in the base of the cup is (as illustrated in Fig. 1.1):

$$\Delta P = P_B - P_T = \rho_w g(h - x) - \rho_w g z \tag{1.1}$$

$$\therefore \Delta P = \rho_w g(h - x - z) \tag{1.2}$$

x is the height of the top of the cup above the water level, and z is the water level inside the cup, as measured from its base. h is the height of the cup. ρ_w is the density of water (about $1,000 \, \text{kgm}^{-3}$) and g is the strength of gravity. $g = 9.81 \, \text{Nkg}^{-1}$, assuming the experiment is performed on Earth. Pressure $P = \rho g h$ results from the idea of the pressure (the *force per unit area*) of a liquid column of height h

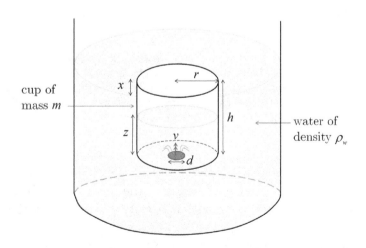

Fig. 1.1   Model of the Election cup, assumed to be a cylinder of height h and radius r, of dry mass m. The cup has a hole of diameter d drilled into the base. x is the distance between the top of the cup and the water line. z is the depth of water inside the cup.

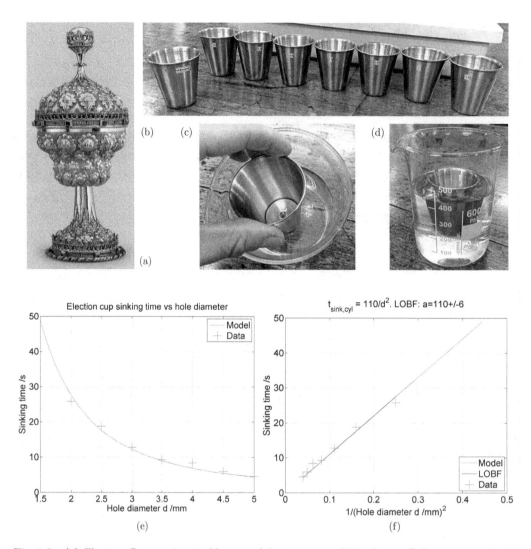

Fig. 1.2   (a) *Election Cups* are inspired by one of the treasures of Winchester College, an ornate 15th century silver-gilt cup. In the *Election Cups* practical, eight identical, but somewhat less expensive cups, illustrated in (b), are provided. Each has a hole drilled in the base varying in diameter from 2.0 mm to 5.0 mm. As shown in (c) and (d), the cups are carefully placed in a beaker of water, and the time to sink is recorded. One can show that the sinking time is inversely proportional to the square of the diameter of the hole, as demonstrated in sub-figures (e) and (f).

and density ρ being $P = \rho g h$. It is the *weight* (the mass $m = \rho A h$ of a column of cross-section A multiplied by the strength of gravity g), divided by the cross-section A. i.e. $P = mg/A = \rho A h g / A = \rho g h$.

Now $\Delta P v$ is the work done on 'one second of water', per unit cross sectional area, travelling through the hole. If resistive losses and any gravitational potential

(a)

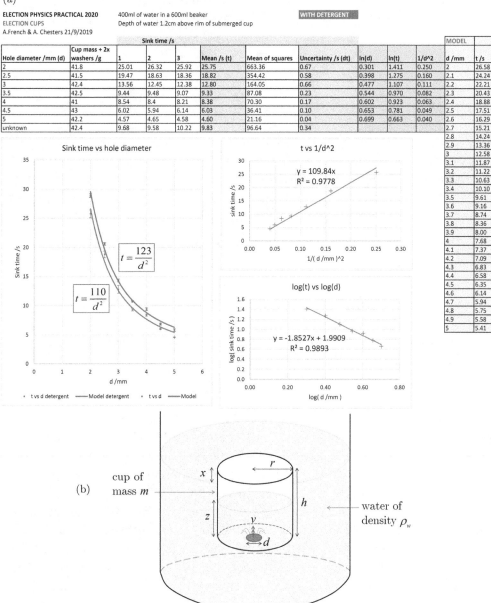

Fig. 1.3 Implementation of a model of *Election Cups* (i.e. sinking time vs hole diameter) and analysis of experimental data using a spreadsheet. A strong correlation between the sinking time and the inverse-square of the hole diameter enables one to predict the unknown diameter of a cup given its sinking time. The spreadsheet and embedded graphs are a clear method of presenting all calculations, including those associated with quantifying uncertainty in experimental measurements.

energy change can be ignored, then this must equate to the kinetic energy gained $\frac{1}{2}mv^2$. The mass m of one second of water is $\rho_w v$ per unit cross-section, hence:

$$\Delta P v = \tfrac{1}{2}\rho_w v \times v^2 \tag{1.3}$$

$$\therefore \ \Delta P = \tfrac{1}{2}\rho_w v^2 \tag{1.4}$$

This is effectively a statement of *Bernoulli's Principle*.[6]

$$\Delta P = \rho_w g(h - x - z) = \tfrac{1}{2}\rho_w v^2 \tag{1.5}$$

This means we can write the water outflow speed in terms of the heights h, x, z:

$$v = \sqrt{2g(h - x - z)} \tag{1.6}$$

Now let $V = \pi r^2 z$ be the water volume in the cylindrical cup. Continuity of water[7] means the rate of change of volume V is:

$$\frac{dV}{dt} = C\pi \tfrac{1}{4}d^2 v \tag{1.7}$$

We can combine the water volume balance $\frac{dV}{dt} = C\pi \tfrac{1}{4}d^2 v$, with cylinder volume $V = \pi r^2 z$, and our energy-balance equation for inflow speed $v = \sqrt{2g(h - x - z)}$:

$$\pi r^2 \frac{dz}{dt} = C\pi \tfrac{1}{4}d^2 \sqrt{2g(h - x - z)} \tag{1.8}$$

The empirical constant C, which will be within range $0 < C < 1$, is an adjustment for the effects of fluid viscosity, resulting in a slower flow rate of water into the cup than predicted. A *vena contracta* of $C \approx 0.65$, i.e. a diametric contraction of fluid flow into the hole by 35% (Nakamura & Boucher [30], p. 67) is predicted.

Newton's second law of motion equates the **mass $\times$ acceleration** of the cup to the **vector sum of forces upon it**. These forces are: upthrust from the water surrounding the submerged cup, and the weight of the cup + the water within it. We shall assume other possible forces such as drag and surface tension are negligible by comparison.[8] *Archimedes' principle* of buoyancy states the upthrust is the *weight of fluid displaced by the cup*. Now let us make a bold simplification, and assume the

---

[6] *Bernoulli's Principle*: For low speed, *incompressible* flow, pressure p and fluid speed v are related by: $p + \tfrac{1}{2}\rho v^2 + \rho g z =$ constant. ρ is the density of the fluid, g is the strength of gravity and z is a height above a particular gravitational potential reference. In our *Election Cups* example we can ignore changes in z over the hole, so the pressure difference is $\Delta P \approx \tfrac{1}{2}\rho v^2$.

[7] *Continuity of water* means "mass flow in equals mass flow out." For most sensible human scenarios, water is essentially incompressible, so continuity means the input volume flow-rate must equal the output volume flow-rate. Call this the First Law of Plumbing.

[8] We can reduce the effect of surface tension by adding a few drops of detergent.

cup sinks in such a stately fashion that *acceleration can be neglected*. This means a form of *dynamic equilibrium* where the upthrust equals the weight:

$$\underbrace{\rho_w \pi r^2 (h-x)g}_{\text{weight of fluid displaced}} = \underbrace{mg + \rho_w \pi r^2 zg}_{\text{weight of cup + water in it}} \qquad (1.9)$$

$$\implies h - x - z = \frac{m}{\rho_w \pi r^2} \qquad (1.10)$$

Hence by substitution $h - x - z = \frac{m}{\rho_w \pi r^2}$ into $\pi r^2 \frac{dz}{dt} = C\pi \frac{1}{4}d^2 \sqrt{2g(h-x-z)}$:

$$r^2 \frac{dz}{dt} = C\frac{1}{4}d^2 \sqrt{\frac{2mg}{\rho_w \pi r^2}} \qquad (1.11)$$

We now have a differential equation for the water level $z(t)$ in the cup which can be solved very simply by integration. By dividing both sides by r^2, you can see that $\frac{dz}{dt}$ is time independent, or alternatively, the gradient of the z vs t graph is a constant.

$$\int_0^z dz' = \frac{1}{4}\frac{Cd^2}{r^2} \sqrt{\frac{2mg}{\rho_w \pi r^2}} \int_0^t dt' \qquad (1.12)$$

$$\therefore z(t) = \frac{1}{4}\frac{Cd^2}{r^3} \sqrt{\frac{2mg}{\rho_w \pi}}t \qquad (1.13)$$

Now $h - x - z = \frac{m}{\rho_w \pi r^2} \implies x = h - \frac{m}{\rho_w \pi r^2} - z$, hence:

$$x(t) = h - \frac{m}{\rho_w \pi r^2} - \frac{1}{4}\frac{Cd^2}{r^3}\sqrt{\frac{2mg}{\rho_w \pi}}t \qquad (1.14)$$

Also: $V = \pi r^2 z$, so:

$$V(t) = \frac{1}{4}\pi C\frac{d^2}{r}\sqrt{\frac{2mg}{\rho_w \pi}}t \qquad (1.15)$$

In other words, we expect quantities z, x, V to all vary linearly with time. The cup sinks at time t_{sink} when the water level to cup rim displacement $x = 0$, i.e. when the cup is completely submerged. $\therefore z_{\text{sink}} = h - \frac{m}{\rho_w \pi r^2}$. Hence:

$$t_{\text{sink}} = \left(h - \frac{m}{\rho_w \pi r^2}\right)\frac{4r^3}{Cd^2}\sqrt{\frac{\rho_w \pi}{2mg}} \qquad (1.16)$$

Which means $t_{\text{sink}} \propto 1/d^2$. The equation above defines the constant of proportionality a:

$$t_{\text{sink}} = \frac{a}{d^2} \qquad (1.17)$$

$$a = \left(h - \frac{m}{\rho_w \pi r^2}\right)\frac{4r^3}{C}\sqrt{\frac{\rho_w \pi}{2mg}} \qquad (1.18)$$

$t_{\text{sink}} = \frac{a}{d^2}$ means if we plot t_{sink} vs $1/d^2$ we should expect a *straight line graph through the origin*, of gradient a. Alternatively, since $\log t_{\text{sink}} = \log a - 2 \log d$, this means a graph of $y = \log t_{\text{sink}}$ vs $x = \log d$ should be a straight line of gradient -2, with y intercept being $\log a$. These predicted relationships can be tested experimentally, and results are presented in Figs. 1.2(d), 1.2(e) and 1.3(a). If all other parameters are measured, the gradient a can be used to calculate the *vena contracta* coefficient C, which would be difficult to measure directly.

$$C = \left(h - \frac{m}{\rho_w \pi r^2} \right) \frac{4r^3}{a} \sqrt{\frac{\rho_w \pi}{2mg}} \tag{1.19}$$

The graph of t_{sink} vs $1/d^2$ can also be used to calculate the unknown hole diameter, since if we know t_{sink} we can read off d from the graph. Alternatively we can use algebra. If we know a:

$$d = \sqrt{\frac{a}{t_{\text{sink}}}} \tag{1.20}$$

1.3.3 *Analysis of Election Cups*

An analysis of the Election Cups experiment is illustrated in Figs. 1.2 and 1.3. The cup parameters are: $h = 54\,\text{mm}$, $r = 20.7\,\text{mm}$, $m = 42\,\text{g}$, $\rho_w = 1{,}000\,\text{kgm}^{-3} = 10^3 \times 10^3\,\text{g} \times (10^3\,\text{mm})^{-3} = 10^{-3}\,\text{gmm}^3$. Note since mm, s, g are the natural units for the system, the strength of gravity is defined to be $g = 9.81\,\text{ms}^{-2} = 9.81 \times 10^3\,\text{mms}^{-2}$. When some detergent is added to reduce the effects of surface tension:

$$t_{\text{sink}} = \frac{110 \pm 6}{d^2} \tag{1.21}$$

where t_{sink} is in seconds and hole diameter d is in mm. $a = (110 \pm 6)\ \text{mm}^2\text{s}$ is obtained from a line-of-best fit of the form $y = ax$ were $y = t_{\text{sink}}$ and $x = 1/d^2$. The formula for a in $y = ax$, and its associated error, is stated and derived in Appendix A. The *product moment correlation coefficient*[9] is $R = 0.995$, which demonstrates a strong positive correlation. The unknown hole results in a sinking time of $t_{\text{sink}} = (9.83 \pm 0.34)$ seconds, which means using the formula $t_{\text{sink}} = \frac{110 \pm 6}{d^2}$, we can calculate the hole diameter to be within the range: $\sqrt{\frac{110-6}{9.83+0.34}} \leq d \leq \sqrt{\frac{110+6}{9.83-0.34}} \Rightarrow 3.20 \leq d \leq 3.50$ (mm), i.e. the unknown hole diameter is about $3.35\,\text{mm}$. Using $a = 110\,\text{mm}^2\text{s}$, the *vena contracta* coefficient $C = \left(h - \frac{m}{\rho_w \pi r^2} \right) \frac{4r^3}{a} \sqrt{\frac{\rho_w \pi}{2mg}} \approx 0.45$. This is about 35% smaller than the 0.65 quoted in Nakamura & Boucher [30]. In order to ensure the cups didn't invert before sinking, a few small washers (thin cylindrical discs with large holes) were added to the cup. The washers were chosen to fit snugly into

---

[9]Note there is a slight discrepancy between the R^2 value in Excel for a $y = ax$ line-of-best-fit and the value I have calculated. In Fig. 1.3, it is 0.9778, whereas it should be $0.9946^2 = 0.9892$. I think this is a well-known Excel 'feature' in older versions of the software.

the base of the cup and had large enough holes to not obscure those drilled into the cups. The washer masses were of course included in the $m = 42\,\mathrm{g}$ cup mass! However, it is possible some additional disruption to the flow resulted from the washers, as indicated by the reduction of C. Another obvious discrepancy from the model is that the cups are clearly not perfectly cylindrical. They have a slightly pitched shot-glass design and therefore are truncated cones rather than cylinders.

It is clear from Fig. 1.3 that adding some detergent *reduces* the sinking time. When no detergent is added:

$$t_{\mathrm{sink}} = \frac{123 \pm 7}{d^2} \tag{1.22}$$

The unknown hole results in a sinking time of $t_{\mathrm{sink}} = (11.31 \pm 0.28)$. Hence $\sqrt{\frac{123-7}{11.31+0.28}} \leq d \leq \sqrt{\frac{123+7}{11.31-0.28}} \Rightarrow 3.16 \leq d \leq 3.43$. The ranges of d for detergent and non-detergent cases overlap, indicating plausible agreement.

The analysis performed appears to indicate a positive correlation between model and experiment, as quantitatively evidenced by the $R = 0.995$ product moment correlation coefficient. Note this degree of correlation is calculated by firstly asserting the model has indeed the t_{sink} vs $1/d^2$ dependence. If instead $y = \log t_{\mathrm{sink}}$ vs $x = \log d$ is plotted, as in Fig. 1.3, the power is calculated to be 1.85 rather than 2.00. Does this undermine the efficacy of the model? This is perhaps an interesting discussion point regarding the linearization (and parameter estimation) methods used in analysis of experimental data. In our case we have good physics modelling reason to suspect a $1/d^2$ dependence, and therefore performing a line of best fit using t_{sink} vs $1/d^2$ is a sensible approach. The line of best-fit yields the constant of proportionality and hence allows us to calculate the *vena contracta* coefficient C from the data. This single parameter scales an idealized scenario and incorporates various real-world deviations such as the effect of fluid viscosity, but importantly, is deemed universal for our system. If instead we tried to find the optimal C value for *each* hole diameter, we would only gain more information on the basis that our $1/d^2$ model is deemed true from the outset. Using the assumed truth of the model in an assessment of the model itself is a circular argument, and should be avoided as dodgy Science.

To summarize: in our case t_{sink} vs $1/d^2$ dependence is evidenced by the $R = 0.995$ correlation between measured sinking time and measured $1/d^2$. The gradient $a = 110 \pm 6$ of t_{sink} vs $1/d^2$ yields a (single) parameter of the system, which we can (using our modelling) then use to calculate the *vena contracta* coefficient C.

The alternative approach of fitting a more general $t_{sink} = bd^n$ model, and therefore $\log t_{sink} = \log b + n \log d$, is pure empiricism and doesn't really link to any underlying theory. It also requires two parameters, rather than one, to be calculated, and both b and n will be subject to uncertainty. Hence from an analysis perspective, the t_{sink} vs $1/d^2$ approach is preferable. Of course the most direct comparison of model and data is simply to plot the measured t_{sink} times vs the model (i.e. predicted) t_{sink} times, using the full range of d values. Then perform a $y = mx$ fit, and investigate the deviation from the ideal $y = x$ outcome. However, this requires all the parameters of the model (a in our case of t_{sink} vs $\frac{a}{d^2}$) to be calculated first. A compromise might be for the experiment described here to be used to compute C, and then a series of repeat experiments performed using exactly the same equipment. In this situation we could assess how close a graph of measured vs model sinking times correlates with a $y = x$ direct proportion.

1.4 Snails of Pursuit

1.4.1 *Context*

This is a very cute modelling example developed from Gnädig, Honyek and Riley's book of *200 Puzzling Physics Problems* [18], and is aimed as a contrast to *Election Cups*. The latter is very much anchored in the slightly messy real-world, and incorporates some empiricism (the *vena contracta* parameter C) and a number of approximations (e.g. a neglect of acceleration of the cup) in the genesis of the model of sinking time vs hole diameter. In *Snails of Pursuit* we dispense with the requirement to align with a realistic experiment, and instead focus on how the model system itself operates and how the various associated parameters scale with each other. This turns out to be rather a beautiful thing.

The idea is that N identical snails are initially positioned at the vertices of a regular N-gon, as illustrated in Fig. 1.4. Each snail, perhaps entranced by powerful gastropod perfume, moves towards the nearest snail in an anti-clockwise fashion. The snails can only slide along at a constant speed u, but can easily change direction. The goal is to work out the subsequent paths of the snails, and calculate how long it takes them to meet at the centre of the N-gon. Although a perhaps a little contrived, *Snails of Pursuit* is nonetheless a powerful demonstration of the application of symmetry ideas, and the good use of an appropriate coordinate system. In this case, *polar coordinates* (range r and anticlockwise angle θ) rather than Cartesians (i.e. x, y at right angles) in the first instance. Plotting the resulting trajectory is also a recipe for creating rather an elegant mathematical art piece, as illustrated in Fig. 1.5.

1.4.2 *Modelling the snails of pursuit*

Snails move towards each other at constant speed u. Since all snails are identical, this means the N-gon shape of their positions will be preserved, but the N-gon will rotate and shrink over time. The position of a neighbouring pair of snails will always form an isosceles triangle with the centre of the N-gon being the third triangle vertex. The angle between the snails is $\alpha = \frac{2\pi}{N}$ and the other two angles of the isoceles triangle are therefore $\beta = \frac{1}{2}(\pi - \alpha)$. This geometric idea is the key to solving the problem of the snail trajectories $r(t)$ and $\theta(t)$ in polar coordinates, as it enables one to simply write down an expression for the radial velocity $\dot{r} = \frac{dr}{dt}$ and tangential

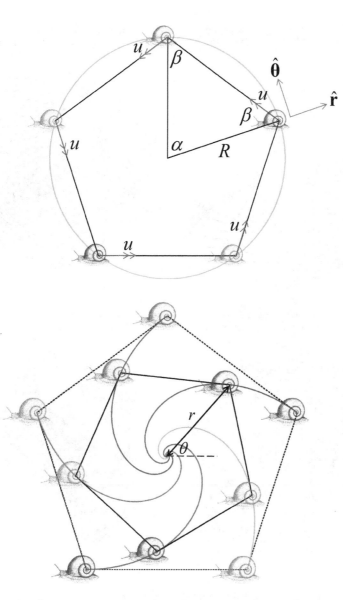

Fig. 1.4 Five *snails of pursuit* are arranged at the vertices of a regular pentagon. The snails can only move at constant speed u, but are free to change direction. If the snails all follow their nearest neighbour, this will result in a shrinking and rotation of the pentagon shape connecting their positions. Snail velocities can be described in terms of radial $\hat{\mathbf{r}}$ and tangential $\hat{\boldsymbol{\theta}}$ polar coordinate unit vectors.

velocity $r\dot{\theta} = r\frac{d\theta}{dt}$. Both these quantities turn out to be *constants*. $r(t)$ can therefore be found via a trivial integration of $\dot{r}$. The polar angle $\theta(r)$, and hence $\theta(t)$, can then be found using a neat Calculus trick $\frac{dr}{d\theta} = \frac{\dot{r}}{\dot{\theta}}$ that is often employed in solving equations of orbital motion.

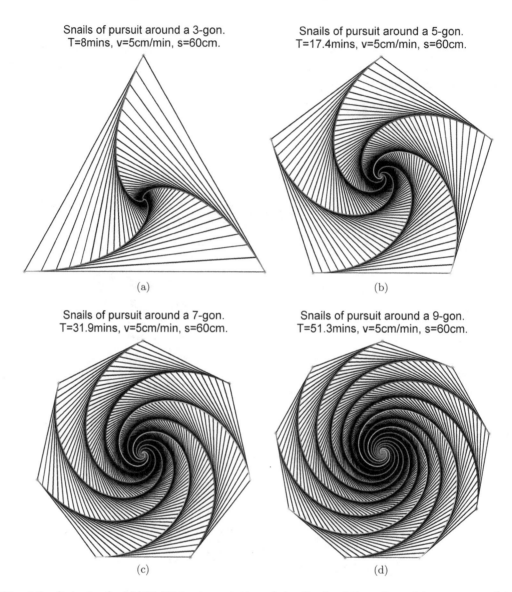

Fig. 1.5 Outputs of a MATLAB implementation of the *Snails of Pursuit* model are presented in subfigures (a)–(d). For each N-gon, the coloured curves represent the inspiralling snail trajectories. A sequence of rotated and scaled N-gons is drawn by connecting the positions of the snails at fixed time intervals. A nice feature of a programming language like MATLAB is that, once programmed, the graphics can be auto-generated, following a change in input parameters such as N. This can include calculations (such as the time to reach the origin) which are presented in the figure title.

To model the dynamics of the snails, consider a polar coordinate system for a particular snail. Using the notation $\dot{r} = dr/dt$, the snail velocity is:

$$\mathbf{v} = \dot{r}\hat{\mathbf{r}} + r\dot{\theta}\,\hat{\boldsymbol{\theta}} \qquad (1.23)$$

where $\hat{\mathbf{r}}$ and $\hat{\boldsymbol{\theta}}$ are the respective mutually perpendicular *unit vectors*[10] in the radial and tangential directions. These unit vectors relate to Cartesian $\hat{\mathbf{x}}, \hat{\mathbf{y}}$ unit vectors by:

$$\hat{\mathbf{r}} = \cos\theta\hat{\mathbf{x}} + \sin\theta\hat{\mathbf{y}} \tag{1.24}$$

$$\hat{\boldsymbol{\theta}} = -\sin\theta\hat{\mathbf{x}} + \cos\theta\hat{\mathbf{y}} \tag{1.25}$$

Note this means:

$$\frac{d\hat{\mathbf{r}}}{dt} = (-\sin\theta\hat{\mathbf{x}} + \cos\theta\hat{\mathbf{y}})\,\dot{\theta} = \dot{\theta}\hat{\boldsymbol{\theta}} \tag{1.26}$$

$$\frac{d\hat{\boldsymbol{\theta}}}{dt} = (-\cos\theta\hat{\mathbf{x}} - \sin\theta\hat{\mathbf{y}})\,\dot{\theta} = -\dot{\theta}\hat{\mathbf{r}} \tag{1.27}$$

Now since the centre-to-neighbouring-snails triangle rotates with the polar coordinate system:

$$\dot{r} = -u\cos\beta \tag{1.28}$$

$$r\dot{\theta} = u\sin\beta \tag{1.29}$$

which means the snail velocity $\mathbf{v} = \dot{r}\hat{\mathbf{r}} + r\dot{\theta}\hat{\boldsymbol{\theta}}$ can be fully determined.

$$\mathbf{v} = -u\cos\beta\hat{\mathbf{r}} + u\sin\beta\hat{\boldsymbol{\theta}} \tag{1.30}$$

If the N-gon bounding the snails has radius R, the radial velocity $\dot{r} = -u\cos\beta$ implies:

$$r(t) = R - ut\cos\beta \tag{1.31}$$

Which means it takes the snail a time:

$$T = \frac{R}{u\cos\beta} \tag{1.32}$$

to reach the centre of the N-gon, and meet the other snails. This is perhaps a surprisingly simple calculation! The tangential velocity $r\dot{\theta} = u\sin\beta$ can be used to find the angular speed of the snail:

$$\dot{\theta}(t) = \frac{u\sin\beta}{R - ut\cos\beta} \tag{1.33}$$

A calculus trick, which is very usefully employed in the analysis of planetary orbits and derivation of Kepler's Laws, can be used to find $r(\theta)$. Since we already know $r(t) = R - ut\cos\beta$, this means we can then find $\theta(t)$ without having to integrate

---

[10]Unit vectors have a magnitude of unity (one), and are used to convey direction without scaling a vector quantity. They are very powerful and elegant when used in an algebraic context. The unit vectors can effectively impose the coordinate system, so you don't need to write things explicitly in component form.

$\dot{\theta}(t) = \frac{u \sin \beta}{R - ut \cos \beta}$ directly.[11] The trick is to firstly write $\frac{dr}{d\theta}$ as a ratio of radial and angular velocities:

$$\dot{r}/\dot{\theta} = \frac{dr}{dt} \bigg/ \frac{d\theta}{dt} = \frac{dr}{d\theta} \tag{1.34}$$

Hence:

$$\frac{dr}{d\theta} = \frac{\dot{r}}{\dot{\theta}} = \frac{-u \cos \beta}{\frac{1}{r} u \sin \beta} \tag{1.35}$$

$$\frac{dr}{d\theta} = -r \cot \beta \tag{1.36}$$

$$\int_R^r \frac{1}{r'} dr' = -\cot \beta \int_{\theta_0}^\theta d\theta' \tag{1.37}$$

$$\ln\left(\frac{r}{R}\right) = -(\theta - \theta_0) \cot \beta \tag{1.38}$$

$$\theta = \theta_0 + \ln\left(\frac{R}{r}\right) \tan \beta \tag{1.39}$$

Therefore:

$$\theta(t) = \theta_0 + \ln\left(\frac{R}{R - ut \cos \beta}\right) \tan \beta \tag{1.40}$$

where θ_0 is the starting angle of the snails. For all the figures in this chapter, the nth snail has starting polar angle: $\theta_0 = \frac{\pi}{2} + (n - 1)\frac{2\pi}{N}$. The relationship $\theta = \theta_0 + \ln\left(\frac{R}{r}\right) \tan \beta$ also explains why the motion of each snail is a *logarithmic spiral*:

$$\theta = \theta_0 + \ln\left(\frac{R}{r}\right) \tan \beta \tag{1.41}$$

$$e^{\frac{\theta - \theta_0}{\tan \beta}} = \frac{R}{r} \tag{1.42}$$

$$\therefore \frac{r(\theta)}{R} = e^{-\frac{\theta - \theta_0}{\tan \beta}} \tag{1.43}$$

The snail acceleration $\mathbf{a}$ can be found from $\mathbf{v} = -u \cos \beta \hat{\mathbf{r}} + u \sin \beta \hat{\boldsymbol{\theta}}$

$$\mathbf{a} = \frac{d\mathbf{v}}{dt} = \frac{d}{dt}\left(-u \cos \beta \hat{\mathbf{r}} + u \sin \beta \hat{\boldsymbol{\theta}}\right) \tag{1.44}$$

$$= -u \cos \beta \frac{d\hat{\mathbf{r}}}{dt} + u \sin \beta \frac{d\hat{\boldsymbol{\theta}}}{dt} \tag{1.45}$$

$$= -u \cos \beta \left(\dot{\theta} \hat{\boldsymbol{\theta}}\right) + u \sin \beta \left(-\dot{\theta} \hat{\mathbf{r}}\right) \tag{1.46}$$

---

[11]Although this is not particularly difficult, using the standard integral result: $\int \frac{f'(t)}{f(t)} dt = \ln|f(t)| + c$.

$$= -u\dot{\theta}\left(\hat{\mathbf{r}}\sin\beta + \hat{\boldsymbol{\theta}}\cos\beta\right) \tag{1.47}$$

$$\therefore \mathbf{a} = -\frac{u^2}{r}\sin\beta\left(\hat{\mathbf{r}}\sin\beta + \hat{\boldsymbol{\theta}}\cos\beta\right) \tag{1.48}$$

The magnitude of snail acceleration $a = |\mathbf{a}|$ is:

$$a(t) = \frac{u^2\sin\beta}{R - ut\cos\beta} \tag{1.49}$$

which asymptotes to an infinitely large value as $t \to T$. Since the radial velocity is constant, this is a result of an ever faster angular acceleration $\ddot{\theta}$. At this point, the snail model of constant speed will no doubt break down, particularly as the snails will now be on top of one another in a slimy writhing mass.

Note the above expression for acceleration $\mathbf{a}$ could also be determined by substituting $\dot{r} = -u\cos\beta$, $\dot{\theta} = \frac{u}{r}\sin\beta$, $\ddot{r} = 0$, and $\ddot{\theta} = -\frac{u}{r^2}\dot{r}\sin\beta = \frac{u^2}{r^2}\cos\beta\sin\beta$ into the general formula for acceleration in polar coordinates:

$$\mathbf{a} = \left(\ddot{r} - r\dot{\theta}^2\right)\hat{\mathbf{r}} + \left(r\ddot{\theta} + 2\dot{r}\dot{\theta}\right)\hat{\boldsymbol{\theta}} \tag{1.50}$$

This acceleration formula yields the familiar *centripetal* $-r\dot{\theta}^2\hat{\mathbf{r}}$ and *tangential* $r\ddot{\theta}\hat{\boldsymbol{\theta}}$ terms of acceleration defined in plane polar coordinates, which indeed are the only terms if $\dot{r} = 0$. The polar coordinate expression for $\mathbf{a}$ can be proven by evaluating the derivative $\mathbf{a} = \frac{d\mathbf{v}}{dt} = \frac{d}{dt}\left(\dot{r}\hat{\mathbf{r}} + r\dot{\theta}\hat{\boldsymbol{\theta}}\right) = \dot{r}\frac{d\hat{\mathbf{r}}}{dt} + \ddot{r}\hat{\mathbf{r}} + r\dot{\theta}\frac{d\hat{\boldsymbol{\theta}}}{dt} + r\ddot{\theta}\hat{\boldsymbol{\theta}} + \dot{r}\dot{\theta}\hat{\boldsymbol{\theta}}$.

Using $\frac{d\hat{\mathbf{r}}}{dt} = \dot{\theta}\hat{\boldsymbol{\theta}}$ and $\frac{d\hat{\boldsymbol{\theta}}}{dt} = -\dot{\theta}\hat{\mathbf{r}}$, this means:

$$\mathbf{a} = \dot{r}\dot{\theta}\hat{\boldsymbol{\theta}} + \ddot{r}\hat{\mathbf{r}} + r\dot{\theta}\left(-\dot{\theta}\hat{\mathbf{r}}\right) + r\ddot{\theta}\hat{\boldsymbol{\theta}} + \dot{r}\dot{\theta}\hat{\boldsymbol{\theta}} \tag{1.51}$$

$$\mathbf{a} = \left(\ddot{r} - r\dot{\theta}^2\right)\hat{\mathbf{r}} + \left(r\ddot{\theta} + 2\dot{r}\dot{\theta}\right)\hat{\boldsymbol{\theta}} \tag{1.52}$$

1.5 Summary

Election Cups and *The Snails of Pursuit* hopefully illustrate the details of what mathematical modelling means in practical terms, and how we shall proceed in future chapters. The language shall be overtly mathematical, and I shall assume familiarity with technical concepts associated with algebra, trigonometry, statistics and calculus that are taught at A-Level in the United Kingdom. However, I will endeavour to explain the meaning of all terms, and indeed the steps of mathematical arguments as much as possible, so that the text can be read from line to line, ideally independent of other sections of the book. I have always been drawn to the details of arguments, particularly those of a mathematical flavour. Alas, the more one experiences, questions of understanding tend to multiply. But ignorance is a delusion of bliss. In *Science by Simulation* I will try to avoid using disingenuous phrases such as "it can be shown," "an exercise to the reader," or worst of all, "it is trivial to prove." These phrases really mean: "the argument is tedious, long or possibly too complex for me to understand, let alone teach." I suggest that all students roll their eyes and give these phrases the disdain they deserve. Or at very least appreciate their true meaning. 'Proof by intimidation' is not a valid approach and one should be honest about what you have mastered, and be cheerfully curious about what remains a mystery. If you don't understand something, and you are told it is trivial or obvious, then politely ask your teacher to demonstrate *why* it is trivial. The truth of the matter will soon become clear, and you will not be wasting their time. For mathematical texts, brevity should always be sought over verbosity, but *clarity is the key goal which trumps all*. I hope I strike an acceptable balance in this book.

Chapter 2

The Epidemiology of Eyam

2.1 Introduction and Context

2.1.1 *The great plague reaches Eyam*

In the autumn of 1665, a damp bale of cloth was delivered from London to the Derbyshire village of Eyam.[1] George Viccars, the tailor's assistant, hung the cloth out to dry in front of his hearth. The evaporating moisture released fleas (*Pulex Irritans*), hungry to replenish their blood-meal stores from human and animal hosts. Unfortunately, the fleas were infected with the stick-shaped *Yersinia Pestis* bacterium [68], (see Fig. 2.2(a)), the microbiological agent of the Great Plague of London, which would eventually cause the death of about a quarter of the population of the capital city within about 18 months [66]. The effect upon the 600–800 pre-plague population of Eyam was similarly devastating. By 1666, over 260 villagers had perished. Contracting plague was a most gruesome affair. Symptoms might start with a flu-like illness and develop into high fever, blood vomiting, lymph node swelling (these are called *buboes*), seizures, decomposition of the body and eventually death.

Multiple transmission pathways of plague have been proposed, with human *ectoparasites* such as fleas or body lice (*Pediculus Humanus Humanus*) currently being the most favoured *vectors*, rather than the 'traditional' hypothesis that plague is spread via contact with rat parasites [6, 31, 58].

The plague bacterium *Yersinia Pestis* is thought to be the source of what history refers to as the *Black Death*, which may have caused the loss of around *half* of Europe's population in the 14th century AD, a catastrophe predating Great Plague

---

[1]The inspiration for this chapter was a study by Dr. John Cullerne of Winchester College of the play *The Roses of Eyam* [53]. In addition to performing the play, his class also discussed the epidemiological models of the Eyam plague. John and I have since collaborated in the production of a pair of *Physics Education* papers [5, 11], that form the backbone of this chapter.

of London by centuries. Untreated *bubonic* plague has a fatality probability of about 50% in humans. The related *pneumonic* plague (in which the infection reaches the lungs, perhaps via inhalation of infectious fluid droplets) will often result in death within three days if untreated [31]. Our modern reliance on antibiotics (which act upon bacterial pathogens such as *Yersinia Pestis,* but *not* viruses) cannot be understated, and a failure to invest adequately in the development of new antibiotic variants, is a crime that our descendants will not forgive.

2.1.2 *William Mompesson's parish records constitute epidemiological data for a closed system*

Reverend William Mompesson of Eyam persuaded the villagers to maintain a state of quarantine,[2] with minimal human contact. Although this did not prevent a huge human tragedy, it did help prevent the spread of the disease to nearby large towns and cities such as Sheffield [62, 67]. During the plague, Mompesson recorded the cumulative death toll, and also the numbers of people who were infected, over a 4-month period. In this chapter, we will use this data to develop a simple epidemiological model of the spread of infection in Eyam, perhaps a rare example of a closed system in an epidemiological sense. Such a model can yield useful insights into the more general behaviour of epidemics, and act as a reality-check for more complex epidemiological models, where the interactions between the *Susceptible S, Infective I, Dead D* and *Recovered R* populations are more subtle and therefore harder to intuit.

Barton [3] presents a summary of Mompesson's records, which are given in the table below. Note these refer to a particular outbreak, rather than the entire record of plague in Eyam. About 100 more deaths, and additional Susceptibles and Infectives, are not accounted for in this snapshot. From a modeling perspective, a longer-term study might break the approximation of a closed system. The person-fractions derive from interpolating between dates, rather than a bizarre quirk of biology. Only statisticians have 2.4 children.

Time (t)	Susceptible (S)	Infective (I)	Dead (D)	Recovered (R)
0.00	235	14.5	0	0
0.51	201	22	26.5	0
1.02	153.5	29	67	0
1.53	121	21	107.5	0
2.04	108	8	133.5	0
2.55	97	8	144.5	0
3.57	83	0	166.5	0

$$(2.1)$$

---

[2]During the events of a plague outbreak, the city-state of Ragusa (modern Dubrovnik) established a rule where newcomers to the city were to be confined to nearby islands for 40 days (initially 30). The Italian translation for '40 days', *quaranta giorni*, would then become the origin of the word *quarantine* [27], used today to refer to the isolation of Infectives to prevent the spread of a disease during an epidemic.

Plotting S, I, D vs t, (see Fig. 2.2) immediately transforms a moving human story to an intriguing mathematical problem. Ask students to describe the shape of the curves (Fig. 2.2(d)) that might be hinted at by the data points, and you will hopefully arrive at something like "the Dead rises to about 170, but less steeply as time goes on." Or: "The curve of Susceptibles decays, quite steeply at first and less so as time progresses." Most interestingly, the curve of Infectives rises to a peak and then drops to zero. To engage with the data, and to set things up for later on, we would encourage students to plot the results themselves, ideally using Excel and/or Python or MATLAB. Two forms of motivation should naturally lead from this work. First: "Can we predict the shape of the curves using equations for how S, I, D vary with time t?" And second: "Are there characteristic timescales associated with the pestilence? Could the same model, but with different parameters, model another disease such as Malaria, Ebola, Influenza, etc.?" We shall consider this latter question by applying the Eyam equations to the 2014 Ebola epidemic in Liberia [59], and finally to the coronavirus pandemic in 2020 [41].

2.1.3 *Context for the introduction of mathematical modelling methods*

The Eyam story can also provide tangible context for the introduction of several important mathematical modelling techniques.

2.1.3.1 *Modelling population flows using coupled differential equations*

Possibly the most basic idea of the science of epidemiology is to quantify the spread of disease by modelling the population flows (see Fig. 2.2) between three distinct groups: 'Susceptibles' (S) (those who have the potential to contract the disease but don't yet have it), 'Infectives' (I) (those who have the disease and have the potential to infect others), 'the Dead' (D) and 'the Recovered' (R). It is assumed that the Dead are swiftly buried or cremated and therefore no-longer represent a source of infection. This might well be an unfounded assumption when disease is rife and societal infrastructure is limited. At Eyam one assumes nobody recovered from the pestilence, so the fourth 'Recovered' (R) category is unoccupied.[3] Based upon reasonable assumptions we can write down a set of *coupled differential equations* which describe how the *rate of change* of these sub-populations vary. Since most physical laws relating to motion are stated in terms of derivatives (e.g. *Newton's Second Law, Fourier's Law of Heat Conduction, The Wave Equation*) this is a very generic methodology.

---

[3] Although clearly William Mompesson himself survived the Black Death as he died in 1709. We might model him as a Susceptible (S) that remained after the number of Infectives (I) became zero. Note in most literature R means 'removed', i.e. Dead or Recovered. In this book, we shall preserve the distinction, particularly since D populations are often quite well documented, but R populations are estimated.

2.1.3.2 *Solving the Eyam equations*

The set of three differential equations which describe the time variation of S, I, D, R populations will be 'solved', i.e. $S(t), I(t), D(t), R(t)$ will be determined explicitly via a variety of methods:

(1) Iterative solution via a fixed timestep 'Euler' method.
(2) Scaling by characteristic parameters relating to population and time to distill the mathematical essence of the equations (it will also simplify them).
(3) Analytic solution via anti-differentiation (i.e. integration). Unfortunately, one relationship, t as a function of $(R + D)$, will be in integral form and cannot be evaluated explicitly.
(4) Approximate solution of the integral expression for t as a function of $(R + D)$, using a truncated polynomial expansion of one of the exponential terms.
(5) Numeric evaluation of $R + D$, and hence solution of the entire system. We will discuss the *Trapezium Rule* and the use of *Cubic Splines* (see Fig. 2.4).
(6) A stochastic interpretation. The population changes during a fixed timestep shall be random integers from a Poisson probability distribution, whose mean is characterised by a reworking of the Eyam equations. This is perhaps a better model of a real epidemic. To assess general trends (and to compare with the deterministic solutions described previously), we can run the stochastic model hundreds or thousands of times, and take an average (see Figs. 2.7–2.9).
(7) Evaluation via a fluid mechanical analogue. A mathematical description is great, and a software simulation makes this better, but a live, kinesthetic demonstration using fluids, pulleys and plumbing to represent the model can add a sense of theatre, and reality, and will draw in the interest of a wider audience that may at first be put off by an overtly mathematical exposition. The Eyam equivalent to Phillip's famous MONIAC is illustrated in Fig. 2.2(b).

2.1.3.3 *Plotting, linearising and parameter estimation*

The *S,I,D* populations can be identified from Mompesson's records and plotted on a graph vs time. As discussed above, it is assumed that $R = 0$, so $D + R = D$ in the analysis below.[4] The shape of these curves motivates a search for a solution to the differential equations, and any characteristic parameters which are germane to the Eyam pestilence. In this case, a full solution requires a hybrid of analytical and numerical calculus techniques, for example using an iterative method based upon the calculation of population changes over a short but finite time interval.

---

[4]For more general systems where recovery is deemed possible, let $D \to R + D$ and let $D = k(R + D)$ where k is the 'mortality fraction'.

The Eyam model turns out to be based upon two[5] key parameters α, β. To find these requires the formation of a *line of best fit* to a (suitable transformation) of Mompesson's records and then an overall comparison of model predictions to actual data, metricated by a *cost* proportional to the *sum of square deviations* (SSD) of model and measurement. The latter is run in a loop, with one of the model parameters varied for each iteration (see Fig. 2.3).

Linearization and *lines-of-best* fit are techniques that underpin the analysis methods of much of experimental science. Before we explore the Eyam system, let us work through an example of a model of a heavy ball being projected vertically, if air resistance can be neglected.

(1) A mathematical model is proposed relating measurable quantities, e.g. vertical displacement x vs time t of a falling ball projected upwards (from $x = 0$) with initial velocity u, with downward gravitational field strength g. If we assume constant acceleration motion, assuming the large mass of the ball allows us to ignore the contribution to acceleration by air resistance, then: $x(t) = ut - \frac{1}{2}gt^2$. This is illustrated in Fig. 2.1.

(2) An algebraic manipulation is then made such that a *straight line (linear) relationship* is predicted, e.g. $\frac{x}{t}$ vs t since $\frac{x}{t} = u - \frac{1}{2}gt$. If a straight line is indeed formed (to appropriate precision), from experimental data, then the overall algebraic structure of the model is deemed a plausible description of real relationships between measurable quantities.

(3) *Model parameters* (e.g. the initial velocity u and the strength of gravity g) can then be obtained from the *intercept* and *gradient* properties of the, suitably extended, straight-line graph. In this case the vertical axis intercept is u and the gradient is $-\frac{1}{2}g$.

In summary, we *linearize* a physical model into a form $y = mx + c$, or if there is good reason to suspect a direct proportion, $y = mx$. If there is a strong linear correlation, then the model is plausible. The *product moment correlation coefficient* r of the line-of-best-fit gives us a metric for how 'good' this fit is. $r = 1$ implies a perfect positive correlation, $r = -1$ implies a perfect negative correlation, and $r = 0$ implies no correlation. In addition, m and c will yield up to two parameters of the model, and from the line of best fit we can also calculate the *uncertainties* in these parameters. See Appendix A for the mathematical details of line-of-best fit calculations, which you can use as a recipe for the implementation of 'linear regression' functionality in computer code.

---

[5] Actually there are *four* key parameters for the Eyam system. In addition to α, β there are the initial Susceptible and Infective populations S_0, I_0. An alternative quartet, $\{t_{max}, I_{max}, \alpha, \eta\}$, based purely upon the Infective curve, will also be used, particularly for the Ebola data. For the numeric method employed to evaluate $t(D + R)$, there is in addition, a finite timestep Δt.

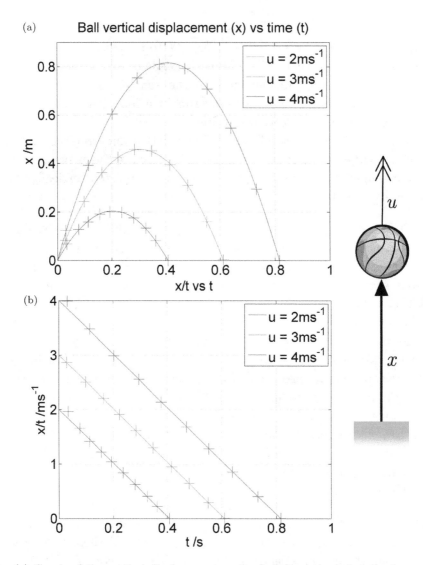

Fig. 2.1   (a) Graph of the vertical displacement x of a ball from the point of release vs time t. It is launched with upward velocity u ms^{-1} and accelerates downwards at g ms^{-2}, i.e. air resistance is deemed negligible. $+$ represents measurements whereas the solid lines describe a constant acceleration model relationship $x = ut - \frac{1}{2}gt^2$. (b) Linearisation of the x vs t variation of a ball falling under gravity. A straight-line relationship between measured values ($+$) implies the underlying model relating x and t is plausible. The gradient and intercepts of the graph can then yield useful model parameters. In this case the intercept yields the initial velocity u and the gradient is $-\frac{1}{2}g$.

2.1.3.4 *Model fitting using a graphical user interface*

The variation of the output of a mathematical model, in our case $S(t), I(t)$, $D(t), R(t)$ with its constituent parameters, is best determined in a dynamic, graphical sense. In other words, a graph whose shape can be animated with a change in parameters. The art of curve sketching from a given equation $y = f(x)$, is a classic test of a student's mettle, particularly at University interviews, and is a superb way to bring many mathematical ideas together to help solve a focused problem (i.e. what shape does the curve have, where are the zeros, stationary points and asymptotes?) However, visualising a general set of curves that change as its parameters are varied is not a static thing, and may be tough to draw by hand with precision, especially if certain geometric features require numeric methods (iteration, numeric integration, etc.) to determine. This is where some computer assistance can be highly useful. In this chapter, and indeed throughout this book, we will employ a bespoke Graphical User Interface (GUI) to the software which runs the model (see Figs. 2.3 and 2.5, for example). The parameters will often be assigned via the movement of a set of sliders, a digital incarnation of a frequency selector in an old analogue radio, or a graphic equaliser in a hi-fi system. The rapid calculating power of the computer enables the model outputs to respond dynamically to the input changes.

From an experimental point of view, a model is only deemed to have validity if it correlates positively with real data. Our software will underlay Mompesson's records with $S(t), I(t), D(t), R(t)$ so a visual comparison can be made. In addition, we shall determine a quantitative metric of fit, based upon the sum of squared differences (SSD) between the model and data. A combination of qualitative visual fit and quantitative measure (i.e. aim to reduce SSD) is the recommended methodology. Although it is possible to perform some form of automated best-fitting algorithm, the visual fit provides a vital sanity check.

2.2 A S, I, D, R Model of the Plague Epidemic in Eyam

2.2.1 *A simple epidemiological model*

A mathematical model to describe how S, I, R, D quantities vary with time is the essence of the applied science of *Epidemiology*. We will endeavour to pose the simplest model that represents the Eyam population dynamics. In real life there are many other aspects to consider. For example, although the population of Eyam was closed in a quarantine sense, we have not factored in births and indeed deaths by non-disease related causes. Also the Dead, particularly if not cremated, may remain a source of infection. We also assume that $R = 0$.

The following model represents perhaps the simplest sensible model of the Eyam pestilence and yields two parameters which might be of general interest to wider epidemiological studies of infection. From an educational point of view, it is also an interesting case study of how a simple iterative numerical model can be used to solve a set of differential equations that describe a system of variables which interact dynamically over a period of time.

The graph in Fig. 2.2 should make the point that the exact mathematical relationship of their curvaceous forms is not easily inferred. However, it is possible to

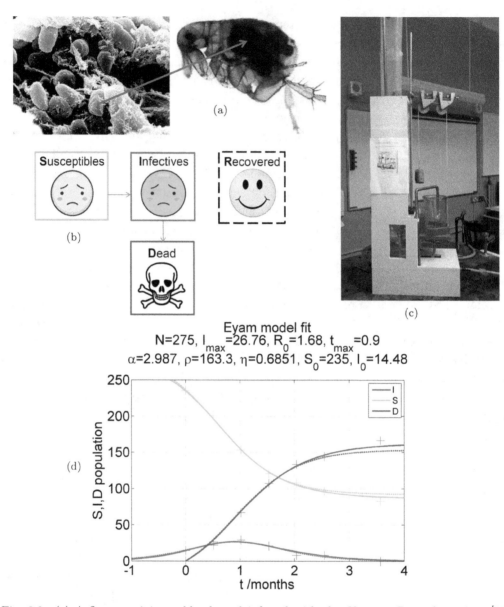

Fig. 2.2 (a) A flea containing a blood meal infected with the *Yersinia Pestis* bacterium [68]. (b) Population flows in the Eyam model. In this rather simple and brutal model, there is no possibility of recovery. Flow is one way from S to I to D. (c) A mechanical/hydrodynamic analogue of the S, I, D equations which comprise the Eyam model. This was inspired by the Phillips Machine or MONIAC (Monetary National Income Analogue Computer) [33] and built by Andy Chesters and John Cullerne of Winchester College. A description of MONIAC is given by Raworth [37]. (d) The variation of number of *Infectives* (I), *Susceptibles* (S) *and Dead* (D) in the village of Eyam between July to October 1666. Mompesson's data ($+$) underlaid with a solution to the Eyam equations. The dashed lines are the approximate 'K&K' curves. The latter will be defined in Section 2.2.8.

make sensible predictions regarding their gradients or 'slopes'. As pointed out earlier, this is how the vast majority of physical laws of motion are posed, from Newtonian to Relativistic, to Quantum variants of Mechanics.

2.2.2 *The Eyam equations*

It makes sense that the rate of change of the number of Dead + Recovered should be proportional to the number of Infectives. Therefore:

$$\frac{d(D+R)}{dt} = \alpha I \tag{2.2}$$

In other words, if we plot the rate of change of the Dead + Recovered vs the number of Infectives, we predict a straight line from the origin of gradient α.

Let the Dead and Recovered populations be a fixed proportion of $D + R$, i.e.

$$D = k(D+R) \tag{2.3}$$

where the 'mortality fraction' is k. For the Eyam plague it is assumed $k = 1$ and therefore $R = 0$. This might prove to be an over-estimate, given $k \approx 0.5$ for untreated plague. However, there does not appear to be any evidence for a Recovered population in Mompesson's parish records. Note if you solve for $R + D$ and know k, you can find D from $D = k(D + R)$ and then R from:

$$R = \frac{1-k}{k} D \tag{2.4}$$

It makes sense that the rate of loss of Susceptibles (S) should be proportional to the number of Infectives (I), and also scale with the number of Susceptibles, i.e. the rate of change of a 1,000 Susceptibles should be 10 times that of a hundred. Hence:

$$\frac{dS}{dt} = -\beta S I \tag{2.5}$$

The parameter β is the (as yet unknown) constant of proportionality, and the minus sign ensures that S decreases as time progresses. If the population is assumed to be fixed, or 'closed', the sum of Dead, Recovered, Susceptible and Infected must equal a constant:

$$I + S + D + R = \text{constant} \tag{2.6}$$

$$\therefore \frac{dI}{dt} + \frac{dS}{dt} + \frac{d(D+R)}{dt} = 0 \tag{2.7}$$

$$\therefore \frac{dI}{dt} = -\frac{dS}{dt} - \frac{d(D+R)}{dt}$$

$$\therefore \frac{dI}{dt} = \beta S I - \alpha I = (\beta S - \alpha) I$$

Our final Eyam equation is therefore:

$$\frac{dI}{dt} = (\beta S - \alpha)I \tag{2.8}$$

2.2.3 *Solution of the Eyam equations via the iterative Euler method*

Alas it is not possible to solve the Eyam system of equations analytically, that is, use a process of *anti-differentiation* to determine exactly how S, I, D, R will vary with time t. However, if we accept the imprecision of using a small but finite time-step, we can use the system of Eyam equations to define an iterative scheme that can readily be evaluated using a spreadsheet or computer program. This approximation of 'step up = slope × step along' is known as the *Euler Method*.[6] Note for brevity, $D + R$ shall be written as D, (i.e. $R = 0$), and this was probably the case anyway for the Eyam plague. The Euler iterative recipe is:

$$t_{n+1} = t_n + \Delta t$$
$$\Delta D_n = \alpha I_n \Delta t$$
$$\Delta S_n = -\beta S_n I_n \Delta t$$
$$\Delta I_n = \beta S_n I_n \Delta t - \alpha I_n \Delta t$$
$$D_{n+1} = D_n + \Delta D_n$$
$$S_{n+1} = S_n + \Delta S_n$$
$$I_{n+1} = I_n + \Delta I_n \tag{2.9}$$

Initial conditions, based upon Mompesson's data are: $D_0 = 0, S_0 = 235$, $I_0 = 14.5$. Based upon the overall duration of the Eyam data (just over 3.5 months), a time-step of $\Delta t = 0.005$ might be sensible to result in visually smooth curves. If a spreadsheet is set up with Δt as a parameter, the effect of reducing (or indeed increasing) it can be readily investigated. The remaining challenge is now to find the parameters that best fit the Mompesson data. This is a powerful motivation to introduce two very useful techniques, one analytic and the other numerical. Although the Eyam differential equations cannot be easily solved as functions of time, it is possible to find out how I varies with S, and this is illustrative of the more general 'separate the variables and then integrate both sides' calculus method:

$$\frac{dS}{dt} = -\beta SI$$

$$\frac{dI}{dt} = \beta SI - \alpha I$$

---

[6]Leonhard Euler 1707–1793, who made enormous contributions to the language and applications of Mathematics, Calculus in particular. Possibly even more than Leibniz and Newton, but let's not wade into that debate...

$$\therefore \frac{dI}{dS} = -\frac{\beta SI - \alpha I}{\beta SI} = \frac{\alpha}{\beta}\frac{1}{S} - 1$$

$$\therefore \int_{I_0}^{I} dI' = \int_{S_0}^{S}\left(\frac{\alpha}{\beta}\frac{1}{S'} - 1\right) dS'$$

$$\therefore I - I_0 = \left[\frac{\alpha}{\beta}\ln S - S\right]_{S_0}^{S}$$

$$\therefore I - I_0 = \frac{\alpha}{\beta}\ln S - \frac{\alpha}{\beta}\ln S_0 - S - S_0$$

$$\therefore \frac{\alpha}{\beta}\underbrace{\ln \frac{S_0}{S}}_{x} = \underbrace{I_0 + S_0 - I - S}_{y} \tag{2.10}$$

$$\therefore y = \frac{\alpha}{\beta}x \tag{2.11}$$

If we use Mompesson's data to plot $y = I_0 + S_0 - I - S$ vs $x = \ln\frac{S_0}{S}$, this should yield a *straight line from the origin* with gradient $\frac{\alpha}{\beta}$. This ratio can be readily obtained via a line-of-best-fit method, either manually using a clear plastic ruler, or via an in-built function in Excel or programmed into Python, MATLAB etc. (see Appendix A). For the Eyam data (see Fig. 2.3(b)):

$$\frac{\alpha}{\beta} \approx 163.3 \tag{2.12}$$

If α can be found separately, the above ratio will yield β. To determine α, the Euler Eyam solver is run for a range of α parameters, and with $\beta = \frac{\alpha}{163.3}$. This is perhaps best implemented in a programming language such as Python or MATLAB rather than Excel. For each iteration, the *Sum of Square Differences* (SSD) between the model and data values of S, I, D are computed and logged.

$$\text{SSD} = \sum_i \left\{(S_{m,i} - S_i)^2 + (I_{m,i} - I_i)^2 + (D_{m,i} - D_i)^2\right\} \tag{2.13}$$

In the equation above, the m suffix denotes the mth model run, evaluated using a unique value of α within range 2.7–3.1. Linear interpolation[7] can be used to evaluate the model S, I, D values which correspond to the times associated with the S, I, D data. The chosen value of α corresponds to the iteration which resulted in the minimum SSD. For the Eyam data (see Fig. 2.3(c)).

$$\alpha \approx 2.99 \tag{2.14}$$

---

[7]That is, draw (possibly different) straight lines between the x, y data points, and evaluate the gaps by using an appropriate line variation $y = mx + c$.

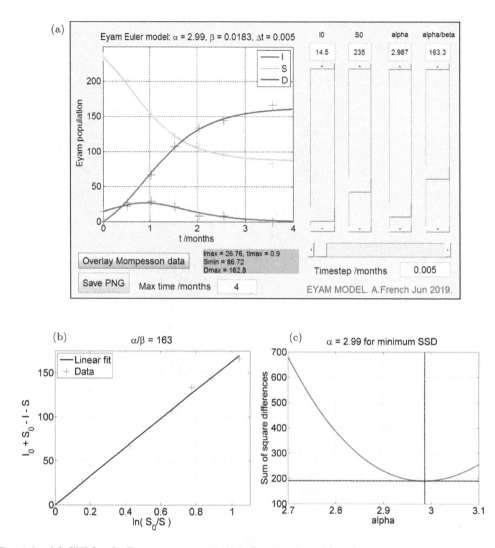

Fig. 2.3   (a) GUI for the Eyam equations model, developed in MATLAB. The four input parameters $I_0, S_0, \alpha, \rho = \frac{\alpha}{\beta}$ can be varied by moving the vertical sliders. The red, green and blue curves (which underlay Mompesson's data) will update dynamically. In this GUI, the Euler method solver is used. The timestep Δt can be controlled using the horizontal slider. (b) Plotting $y = I_0 + S_0 - I - S$ vs $\ln(S_0/S)$ for Mompesson's data enables $\rho = \frac{\alpha}{\beta} \approx 163.3$ to be calculated from the gradient. The Line of Best Fit is determined via linear regression (see Appendix A). (c) Plot of SSD vs α for the Eyam model, evaluated using a Euler method iterative scheme. $\alpha = 2.89$ minimises SSD and hence results in the best fit to the Mompesson data.

Note $\frac{1}{\alpha}$ is a measure of the *time constant* for the Eyam plague. In days it is:

$$\frac{1}{\alpha} = \frac{365}{12} \times \frac{1}{2.99} = 10.2 \tag{2.15}$$

This could be used as a crude measure of the 'infection time', i.e. the approximate number of days from infection till death, or indeed recovery, if we are applying the model more widely and in a contemporary context. If the Eyam model was applied to other, more modern diseases such as Ebola, the α parameter may have a critical relationship to how healthcare is delivered. The smaller $\frac{1}{\alpha}$ is, the greater the likelihood that an Infective may not survive the disease, especially in hard-to-reach rural areas. β might be associated with the contagiousness of the disease, since the higher the number, the greater the rate of change of the number of Susceptibles.

More realistic epidemiological models will almost certainly go beyond the crude parameterisation of α and β, but as a result the models will require extra terms and indeed extra parameters to infer. In their complexity, they may also be a less useful educational vehicle for the introduction of calculus ideas. However, the idea that *all* models are necessarily idealisations, and therefore extensible, is an important point to make to students.

Mompesson's 1666 plague data, underlaid with $S(t), I(t), D(t), R = 0$ model curves via an Euler solver with parameters: $I_0 = 14.5, S_0 = 235, D_0 = 0, \alpha = 2.99, \frac{\alpha}{\beta} = 163.3, \Delta t = 0.005$ is illustrated in graphical form in Fig. 2.2(d), and in MATLAB GUI form in Fig. 2.3(a). This analysis[8] formed the basis of a *Physics Education* paper (2019) "The pedagogical power of context: iterative calculus methods and the epidemiology of Eyam" [11]. The discussion in the next section is adapted from a follow-up *Physics Education* paper (2020) "The pedagogical power of context: extending the epidemiology of Eyam" [5].

2.2.4 *The ups and downs of I*

The Eyam equation for infectives I is:

$$\frac{dI}{dt} = (\beta S - \alpha)I \tag{2.16}$$

It is immediately apparent that $\frac{dI}{dt} = 0$ if $I = 0$ or $S = \frac{\alpha}{\beta}$. By performing a further time derivative, one can see that I is *maximised* when $S = \frac{\alpha}{\beta}$. This is the Susceptible population at the peak of the infection:

$$\frac{d^2 I}{dt^2} = (\beta S - \alpha)\frac{dI}{dt} + I\beta\frac{dS}{dt} \tag{2.17}$$

---

[8]In [11], $\alpha = 2.89$. A refinement of the software used to conduct the analysis (e.g. use of a smaller timestep $\Delta = 0.005$ and more α values in the SSD vs α graph) modifies this to $\alpha = 2.99$.

$$\therefore \frac{d^2 I}{dt^2} = (\beta S - \alpha)^2 I - I^2 \beta^2 S$$

$$\therefore \left(\frac{d^2 I}{dt^2}\right)_{S=\frac{\alpha}{\beta}} = \left(\beta \frac{\alpha}{\beta} - \alpha\right)^2 I - I^2 \beta^2 \frac{\alpha}{\beta}$$

$$\therefore \left(\frac{d^2 I}{dt^2}\right)_{S=\frac{\alpha}{\beta}} = -I^2 \beta \alpha \qquad (2.18)$$

$$\therefore \left(\frac{d^2 I}{dt^2}\right)_{S=\frac{\alpha}{\beta}} < 0 \qquad (2.19)$$

Note we use $\frac{dS}{dt} = -\beta SI$ in the second step. Now since $\frac{\alpha}{\beta}$ is clearly a useful parameter, define:

$$\rho = \frac{\alpha}{\beta} \qquad (2.20)$$

If ρ can be found for a given infection, then this represents a *threshold* for the epidemic. If the 'local susceptible population'[9] is less than this, then the Eyam model predicts the number of infectives will *reduce* rather than grow. Without any further investigation, it is clear that $I(t)$ must form a *single peaked curve*, as long as the initial susceptible population is $> \rho$.

2.2.5 *A problem of initial conditions*

The Euler solver described above requires knowledge of S, I values at $t = 0$, in addition to parameters α, ρ and timestep Δt. D is also fixed to be zero at $t = 0$. Unfortunately this method of modelling is not very applicable to most modern epidemics. For an epidemic such as Ebola in Liberia in 2014 [59], we may only have a record of 'confirmed infectives' vs time, and the assumption of an isolated population with an associated limited susceptible population is likely to be flawed.

However, infectives vs time curves do nonetheless appear to follow similar trends as per the Eyam model. We can therefore flip the logic of the problem: can we fit an Eyam model to work out what the population dynamics would be, given the data? If we can do this, then perhaps we can gain some insight into the size of the susceptible population that would catalyse the epidemic, and then perhaps use this insight to postulate whether certain social practices (e.g. mass social gatherings during funerals following Ebola deaths) are key accelerants of the epidemic. The Eyam equation model is crude compared to most epidemiological systems, but it does have the benefit of being well defined in a mathematical sense. I therefore

---

[9] A modern epidemic might perhaps be modelled as a 'locally closed' population over a suitably truncated timescale.

suggest it may be useful as a reality-check of more complex systems; a sensible first word rather than the last word in analysis terms.

Since S_0 and β are unlikely to be known *a priori*,[10] we shall adopt a more sensible set of initial conditions for the Eyam system, those that were employed by Kendall [24] in his criticism of the seminal (1927) work of Kermack and McKendrick ('K&K') in their *Contribution to the Mathematical Theory of Epidemics* [23]. K&K proposed a time-symmetric-about-the-peak sech^2 model of infectives vs time, whereas Kendall showed that $I(t)$ is more skewed in shape if one considers a longer timescale than the interval associated with the infection peak.

We shall fit a curve based upon the *peak* of the infection, $(t_{\max}, I_{\max})$ and extend time over range $[-\infty, \infty]$ rather than be constrained by some artificial starting point of data collection, as per the Mompesson Eyam 1666 data. We shall also set our Dead count to be *zero* at the *infection peak,* which means a *negative* value prior to this. This might seem a rather perverse thing to do, but it shall turn out to be very useful in simplifying the mathematics of the solution to the Eyam equations. Particularly as $S = \rho$ at the peak of the infection.

Additionally, we shall scale our variables[11]; time by α and S, I, D by ρ. In a similar way that the equations of fluid dynamics can be solved elegantly if characterised by dimensionless parameters groups such as a *Reynolds Number*,[12] we can use this approach to solve the Eyam equations in a more universal fashion.

Our new variables shall be:

$$\tau = \alpha \left(t - t_{\max} \right) \tag{2.21}$$

$$x = \frac{I}{\rho}, \quad y = \frac{S}{\rho}, \quad z = \frac{D + D\left(t = -\infty \right)}{\rho} \tag{2.22}$$

and our initial conditions shall be:

$$\tau = 0; \quad x = x_{\max}; \quad y = 1; \quad z = 0 \tag{2.23}$$

where $x_{\max} = \frac{I_{\max}}{\rho}$.

Note $-D\left(t = -\infty \right)$ is the cumulative dead ($+$ recovered) from the beginning of the model time, to the peak of the infection. So for $z = 0$ at the infection peak, we must subtract $-D\left(t = -\infty \right)$ from D. Why the minus sign?, well $D\left(t = -\infty \right)$ will be *negative*.

---

[10] A sensible guess for α might be easier, given it often seems to be within the range of $2 < \alpha < 4$ months^{-1}. [48, 50, 59].

[11] Don't forget D actually means $D + R$ here.

[12] When a *Reynolds Number* is low enough, a fluid will not behave in a turbulent fashion, regardless of the type of fluid or indeed the geometry and kinematics of the flow. All these factors are packaged into the Reynolds number $\text{Re} = \frac{uL}{\nu}$ where u is the fluid velocity, L is a characteristic linear dimension and ν is the *kinematic viscosity* for the fluid. For a pipe where L relates to diameter or length, the transition from laminar to turbulent flow typically occurs for Reynolds numbers above 2,300 [7].

2.2.6 *Solving the Eyam equations using a 'semi-analytic approach'*

Using our scaled variables, we can re-write our Eyam equations:

$$\frac{dD}{dt} = \alpha I \implies \rho\alpha\frac{dz}{d\tau} = \alpha\rho x \tag{2.24}$$

$$\therefore \frac{dz}{d\tau} = x \tag{2.25}$$

$$\frac{dS}{dt} = -\beta SI \implies \rho\alpha\frac{dy}{d\tau} = -\beta\rho^2 yx \tag{2.26}$$

$$\therefore \frac{\alpha}{\beta}\frac{dy}{d\tau} = -\rho yx \tag{2.27}$$

$$\therefore \frac{dy}{d\tau} = -yx \tag{2.28}$$

$$\frac{dI}{dt} = (\beta S - \alpha)I \implies \rho\alpha\frac{dx}{d\tau} = (\beta\rho y - \alpha)\rho x \tag{2.29}$$

$$\therefore \frac{dx}{d\tau} = \left(\frac{\beta}{\alpha}\rho y - 1\right)x \tag{2.30}$$

$$\therefore \frac{dx}{d\tau} = (y - 1)x \tag{2.31}$$

By this procedure we have distilled the Eyam equations into their most simple form. To solve, firstly divide the first two to find $z(y)$, which means $D(S)$ if we know ρ.

$$\frac{dz}{d\tau} = x, \frac{dy}{d\tau} = -yx \tag{2.32}$$

$$\therefore \frac{dy}{dz} = -y \tag{2.33}$$

$$\therefore \int_1^y \frac{1}{y'}dy' = -\int_0^z dz' \tag{2.34}$$

$$\therefore \ln y = -z \tag{2.35}$$

$$\therefore y = e^{-z} \tag{2.36}$$

The fixed population constraint means:

$$I + S + D = I_{\max} + \rho + 0 \tag{2.37}$$

$$\implies x + y + z = x_{\max} + 1 \tag{2.38}$$

Hence:

$$x = x_{\max} + 1 - e^{-z} - z \tag{2.39}$$

Note that with $\tau = 0, z = 0$, this means that the total population does *not* equal $\rho(x + y + z)$. In fact it is:

$$N = \rho(x + y + z - z_-) \tag{2.40}$$

which means we have a strong incentive to figure out what $z_- = D(t = -\infty)$ is in terms of model inputs. $R_0 = \frac{N}{\rho}$ is called the *Basic Reproduction Number*, which is a very important characteristic of an epidemic, and was eternally present in coronavirus briefings during the 2020 pandemic.

We can use $x = x_{max} + 1 - e^{-z} - z$ and $\frac{dz}{d\tau} = x$ to express an equation for $\tau(z)$. This is an *integral* expression, which unfortunately cannot be evaluated in a closed form sense; it requires a numerical approach.[13] Hence the 'semi-analytic' description of this solution scheme.

$$\frac{dz}{d\tau} = x_{max} + 1 - e^{-z} - z \tag{2.41}$$

$$\therefore \int_0^z \frac{dz'}{x_{max} + 1 - e^{-z'} - z'} = \int_0^\tau d\tau' \tag{2.42}$$

$$\therefore \tau(z) = \int_0^z \frac{dz'}{x_{max} + 1 - e^{-z'} - z'} \tag{2.43}$$

Example graphs of $\tau(z)$ and its integrand $f(z) = (x_{max} + 1 - e^{-z} - z)^{-1}$ are given in Fig. 2.4. To evaluate the integral, a set of *cubic-splines* are firstly fitted to a vector of $\{z, f(z)\}$ values, and then the integral can be (approximately) evaluated by analytic[14] integration of the cubic curve sections. The computer implementation of this method was done in MATLAB, using code adapted from the listing in Hanselman & Littlefield's *Mastering MATLAB 6* [19]. It is clear from the form of $f(z)$, that zero values of the denominator will cause an infinite asymptote. The meaning of this limit is more than pure mathematical consequence, since $\tau(z) = \int_0^z \frac{dz'}{x(z')}$. $x = 0$ means $I = 0$, which represents the boundary values of the infection at times $[-\infty, \infty]$. Since $I \geq 0$, this means the limits of z are the solutions to:

$$x_{max} + 1 - e^{-z_-} - z_- = 0 \tag{2.44}$$

$$x_{max} + 1 - e^{-z_+} - z_+ = 0 \tag{2.45}$$

Since $x(z)$ has a single maximum at $z = 0$, an iterative numeric solution can readily be found by using the efficiently converging *Newton–Raphson*[15] method, starting

---

[13]For example a method which divides up the area under the integrand, approximately, into strips of precisely known size (e.g. trapeziums, parabolae, cubics, etc.) and then sums these. The thinner the strip width, the more precise the numeric integral.

[14]An *analytic* evaluation means without approximation. The integral of a cubic polynomial $y = Ax^3 + Bx^2 + Cx + D$ is: $\int y \, dx = \frac{1}{4}Ax^4 + \frac{1}{3}Bx^3 + \frac{1}{2}Cx^2 + Dx + E$ where E is a constant.

[15]$f(z) = 0$ can be solved by performing the iteration $z_{n+1} = z_n - \frac{f(z_n)}{f'(z_n)}$. This can converge very rapidly as long as you don't encounter a *stationary point*, i.e. where $f'(z) = 0$.

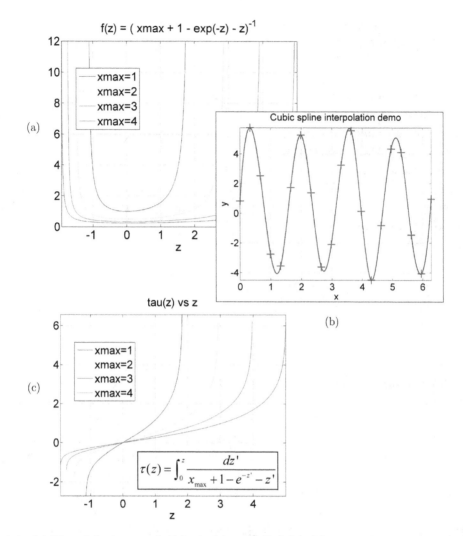

Fig. 2.4 (a) Plot of the integrand $f(z)$ of $\tau(z) = \int_0^z f(z')dz'$. (b) Demonstration of interpolation via cubic splines using MATLAB. The fitting of a piecewise cubic (with continuous first derivatives between cubics) is not just useful for smoothing data. The cubics can also be readily differentiated or integrated, and hence enable a calculus method to be applied to a set of x, y data points. (c) Plots of $\tau(z)$ for various values of $x_{\max}$. The integral has been evaluated numerically via the use of a cubic spline fit between 1,000 equally spaced values between z_- and z_+.

with initial values $z_\pm$.

$$z_\pm^{(n+1)} = z_\pm^{(n)} - \frac{x_{\max} + 1 - e^{-z_\pm^{(n)}} - z_\pm^{(n)}}{e^{-z_\pm^{(n)}} - 1} \tag{2.46}$$

2.2.7 *Eyam equation solution summary*

The method above defines a 'recipe' for the solution of the Eyam equations. In summary:

(1) Input parameters: $\rho, I_{\max}, \alpha, t_{\max}$.

(2) Define: $x_{\max} = \frac{I_{\max}}{\rho}$.

(3) Find limits of z, $[z_-, z_+]$, by solving the following Newton–Raphson iteration:

$$z_\pm^{(n+1)} = z_\pm^{(n)} - \frac{x_{\max}+1-e^{-z_\pm^{(n)}}-z_\pm^{(n)}}{e^{-z_\pm^{(n)}}-1}.$$

Start with $z_\pm^{(0)} = \pm 1$, which avoids any stationary points in the Newton–Raphson iteration, and hence infinite values of $\tau(z)$.

(4) Evaluate numerically the $\tau(z) = \int_0^z \frac{dz'}{x_{\max}+1-e^{-z'}-z'}$ integral, over an equally spaced range between $[z_-, z_+]$. Methods could be the Trapezium rule, 'cubic spline fit and integrate', etc.

(5) Determine $x = x_{\max} + 1 - e^{-z} - z$, $y = e^{-z}$.

(6) Determine S, I, D, t from:

$$t = \frac{\tau}{\alpha} + t_{\max}, \quad I = \rho x, \quad S = \rho y, \quad D = \rho(z - z_-) \tag{2.47}$$

At this point, you may wish to use interpolation to find $D_0 = D(0)$, and then subtract this value from D to set $D = 0$ at $t = 0$. This will enable a direct comparison to the Mompesson data, for example. Note $N = \rho(x + y + z - z_-)$ is still the total susceptible population and will be unchanged by this shift. The only consequence is that the D count will be negative for $t < 0$. The zeroing of D is an option in the Eyam solver GUI illustrated in Fig. 2.5.

(7) D is actually $D+R$. So if you know the mortality fraction k, find $D = k(R+D)$ and $R = \frac{1-k}{k}D$.

(8) Calculate population size N and hence Basic Reproduction Number R_0

$$N = I_{\max} + \rho - \rho z_- \tag{2.48}$$

$$R_0 = \frac{N}{\rho} \tag{2.49}$$

2.2.8 *The K&K approximate solution to the Eyam equations*

2.2.8.1 *The K&K approximation of $\tau(z)$*

The time-symmetric-about-the-peak sech^2 model of infectives vs time $I(t)$ curve proposed by Kermack and McKendrick [23] in the 1920s was orthodoxy until the criticism of Kendall [24] in 1956. As Kendall pointed out, and we will demonstrate using the 2020 UK COVID-19, Liberia Ebola 2014 and the Mompesson Eyam 1966 plague datasets, a symmetric form is not an accurate portrayal of the curve, and indeed can lead to additional errors in the prediction of $D(t)$ and $S(t)$ curves.

The K&K approximation is to assume:

$$e^{-z} \approx 1 - z + \tfrac{1}{2}z^2 \tag{2.50}$$

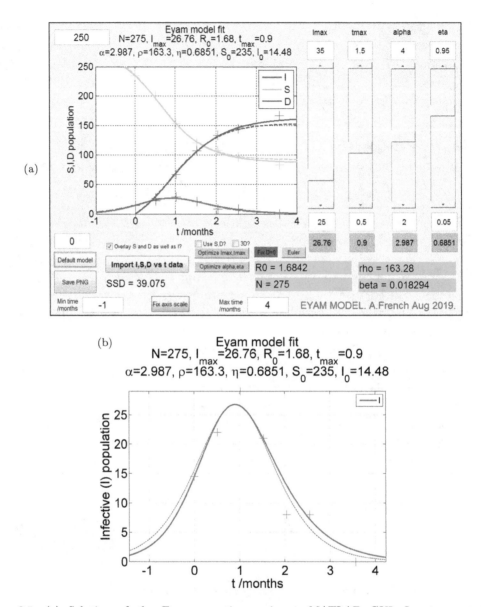

Fig. 2.5 (a) Solution of the Eyam equations using a MATLAB GUI. Input parameters $\{t_{max}, I_{max}, \alpha, \eta\}$ can be varied using the vertical sliders. η is the ratio between the total dead and the total population N, over the extended timescale of the idealised epidemic, $-\infty < t < \infty$. In this example, the model is parameterised to fit Mompesson's 1666 plague data $(+)$, and $D(0) = 0$ is fixed for the model. Dashed lines are the approximate K&K solution. (b) The slight asymmetry of the exact $I(t)$ curve can be seen by comparison to the symmetric K&K sech^2 curve approximation.

which means:

$$x \approx x_{\max} + 1 - \left(1 - z + \tfrac{1}{2}z^2\right) - z \tag{2.51}$$

$$\therefore x \approx x_{\max} - \tfrac{1}{2}z^2 \tag{2.52}$$

For this to be valid, $z \ll 1$, which means the ratio of $D + R$ to the threshold ρ must be smaller than unity, since $z = \frac{D + D(t = -\infty)}{\rho}$. We can immediately see from Fig. 2.4 that the z limits can be much larger than unity as we pass the infection peak, so this approximation will become increasingly crude as time beyond peak infection increases. However, if only for the educational and historical value, let us persist with the analysis.

Under the K&K approximation, the $\tau(z)$ integral becomes:

$$\tau(z) = \int_0^z \frac{dz'}{x_{\max} + 1 - e^{-z'} - z'} \tag{2.53}$$

$$\tau(z) \approx \int_0^z \frac{dz'}{x_{\max} + 1 - \left(1 - z' + \tfrac{1}{2}z'^2\right) - z'} \tag{2.54}$$

$$= \int_0^z \frac{dz'}{x_{\max} - \tfrac{1}{2}z'^2} \tag{2.55}$$

$$= 2 \int_0^z \frac{dz'}{2x_{\max} - z'^2} \tag{2.56}$$

$$= 2 \int_0^z \frac{dz'}{\left(\sqrt{2x_{\max}}\right)^2 - z'^2} \tag{2.57}$$

$$= \frac{2}{\sqrt{2x_{\max}}} \tanh^{-1}\left(\frac{z}{\sqrt{2x_{\max}}}\right) \tag{2.58}$$

where the final step makes use of the standard integral:

$$\int \frac{1}{a^2 - x^2} dx = \tfrac{1}{a} \tanh^{-1}\left(\tfrac{1}{a}x\right) \tag{2.59}$$

Hence:

$$z \approx \sqrt{2x_{\max}} \tanh\left(\sqrt{\tfrac{1}{2}x_{\max}}\tau\right) \tag{2.60}$$

and therefore:

$$x \approx x_{\max} - \tfrac{1}{2}z^2 \tag{2.61}$$

$$= x_{\max} - \tfrac{1}{2}2x_{\max} \tanh^2\left(\sqrt{\tfrac{1}{2}x_{\max}}\tau\right) \tag{2.62}$$

$$= x_{\max} \left\{1 - \tanh^2\left(\sqrt{\tfrac{1}{2}x_{\max}}\tau\right)\right\} \tag{2.63}$$

$$\therefore x \approx x_{\max}\text{sech}^2\left(\sqrt{\tfrac{1}{2}x_{\max}}\tau\right) \tag{2.64}$$

In summary, the K&K approximate solution to the Eyam equations (in terms of scaled variables) is:

$$x \approx x_{\max}\text{sech}^2\left(\sqrt{\tfrac{1}{2}x_{\max}}\tau\right) \tag{2.65}$$

$$z \approx \sqrt{2x_{\max}}\tanh\left(\sqrt{\tfrac{1}{2}x_{\max}}\tau\right) \tag{2.66}$$

$$y = e^{-z} \tag{2.67}$$

Now:

$$\lim_{x\to\pm\infty}\{\tanh x\} = \pm1 \tag{2.68}$$

So, therefore,

$$z_\pm \approx \pm\sqrt{2x_{\max}} \tag{2.69}$$

Hence in the K&K approximation:

$$t = \frac{\tau}{\alpha} + t_{\max}, \; I = \rho x, \; S = \rho y \tag{2.70}$$

$$D + R = \rho\left(z + \sqrt{2x_{\max}}\right) \tag{2.71}$$

$$D = k(D + R) \tag{2.72}$$

$$R = \frac{1-k}{k}D \tag{2.73}$$

$$N = I_{\max} + \rho - \rho\sqrt{2x_{\max}} \tag{2.74}$$

2.2.8.2 *Evaluation of the K&K model*

To assess the efficacy of the K&K approximation, let us firstly apply both 'semi-analytic Eyam' and K&K models to Mompesson's 1666 plague data. The same input parameters are used: $I_{\max} = 26.76$, $t_{\max} = 0.90$, $\alpha = 2.99$, $\rho = 163.3$. To evaluate over a sensible time range (i.e. not infinite!), z limits of $[z_- + \delta z, z_+ - \delta z]$ are used, and for the semi-analytic model, 1000 equally spaced data points are defined over this range. $\delta z = 0.01$ in all simulations described here. The corresponding time range enables the K&K model to be evaluated, again using 1000 equally spaced points.[16] It is worth noting again that the Mompesson data is based upon $D = 0$

---

[16]Note an equally spaced z vector will not mean an equally spaced τ vector of values, since these quantities are not proportional. However, use of 1,000 samples should mean any variation in spacing intervals can be largely neglected on the scale that plots are constructed. With this density of data points, all curves should appear visually smooth.

at $t = 0$. Initial S, I values are $S(0) = 235$, $I(0) = 14.5$, giving a total population of 250. Although when to start the dead count is arbitrary, the total associated population is not, as we have shown in previous sections. In reality, it is likely there would have been deaths due to plague at Eyam before Mompesson started collecting data. Calculating N using the semi-analytic procedure yields $N = 275$, so about 25 people would have died of plague at Eyam before Mompesson's parish records began, had the infection obeyed the idealised Kendall curve from $t = -\infty$.[17] However, to directly compare to Mompesson's data, we shall shift both exact and K&K curves by $D(0)$, so all pass through the origin.

2.2.8.3 *Problems with the K&K approach*

The time symmetric K&K $I(t)$ is not universally true, and the skew of the curve can lead to a discrepancy in predictions, as illustrated in Fig. 2.5. In addition, we also have the problem of what values to assign ρ and α. This is a problem for our semi-analytic method too. Based on field data such as a World Health Organisation (WHO) Situational Report [59], all we have is a set of $\{t, I\}$ data. For the Mompesson data, we showed in the previous section how α, β (and hence $\rho = \frac{\alpha}{\beta}$) could be obtained from the $S(t), I(t), D(t)$ dataset. For most epidemiological scenarios, we don't have this luxury. These problems shall be addressed in Section 2.2.9.

2.2.9 *A semi-analytic solution to the Eyam equations with η instead of ρ*

2.2.9.1 *The problem with ρ*

The Eyam equation solution summary in Section 2.2.7 requires the inputs: $t_{\max}, I_{\max}, \alpha, \rho$. If the only data available is $\{t, I\}$, then $(t_{\max}, I_{\max})$ can be readily guessed directly from the data.[18] The range of α is relatively well bounded in epidemiological literature. For many epidemics, it is typically within the range of $2 < \alpha < 4$, in units of months^{-1} [48–50]. The problem parameter is ρ, which can vary significantly with the populations involved. For Mompesson's 1666 Plague data, $\rho = 163.3$. For Ebola in Liberia in 2014, which will be discussed later, $\rho = 1,538$.

2.2.9.2 *Replacing with ρ with η*

We can't avoid having four inputs to our Eyam model, but we can replace ρ with a different variable that has a much more well-defined range of values. In fact it

---

[17]Actually there was a previous Plague outbreak at Eyam, so the $I(t)$ curve is only valid for the time range associated with Mompesson's data.

[18]Or indeed a good first guess. An initial guess based upon the largest value of I in the data is the default used in the Eyam GUI software described here. This can then be refined automatically, or manually using the GUI sliders.

is possible to find a parameter which has bounds of $[0, 1]$. The total dead (plus recovered) is given by:

$$D_{\text{tot}} = \rho\left(z_+ - z_-\right) \tag{2.75}$$

Clearly this cannot exceed the total population N, or indeed be less than zero, i.e.

$$0 \leq \frac{D_{\text{tot}}}{N} \leq 1 \tag{2.76}$$

$$\therefore 0 \leq \frac{\rho\left(z_+ - z_-\right)}{N} \leq 1 \tag{2.77}$$

$$\therefore 0 \leq \frac{z_+ - z_-}{N/\rho} \leq 1 \tag{2.78}$$

Define a new parameter $\eta = \frac{z_+ - z_-}{N/\rho}$ which *must* be in the range $[0, 1]$. Now since $x = x_{\max} + 1 - e^{-z} - z$, when $x = 0$:

$$x_{\max} + 1 - e^{-z_-} - z_- = 0 \tag{2.79}$$

$$x_{\max} + 1 - e^{-z_+} - z_+ = 0 \tag{2.80}$$

Now the total population $N = \rho\left(x_{\max} + 1 - z_-\right)$. Therefore using $x_{\max} + 1 - e^{-z_-} - z_- = 0$

$$\frac{N}{\rho} - e^{-z_-} = 0 \tag{2.81}$$

$$\therefore z_- = -\ln\left(\frac{N}{\rho}\right) \tag{2.82}$$

From the definition of η:

$$z_+ = \eta\frac{N}{\rho} + z_- \tag{2.83}$$

$$\therefore z_+ = \eta\frac{N}{\rho} - \ln\left(\frac{N}{\rho}\right) \tag{2.84}$$

We can now substitute this in the equation for $x = 0$ for positive z:

$$x_{\max} + 1 - e^{-z_+} - z_+ = 0 \tag{2.85}$$

$$x_{\max} + 1 - z_- - e^{-z_+} - z_+ = -z_- \tag{2.86}$$

$$\underbrace{x_{\max} + 1 - z_-}_{N/\rho} - e^{-z_+} = z_+ - z_- \tag{2.87}$$

$$\frac{N}{\rho} - e^{-\eta\frac{N}{\rho} + \ln\left(\frac{N}{\rho}\right)} = \frac{N}{\rho}\eta \tag{2.88}$$

With the last step noting:

$$\eta = \frac{z_+ - z_-}{N/\rho} \tag{2.89}$$

$$\therefore \frac{N\eta}{\rho} = z_+ - z_- \tag{2.90}$$

Hence:

$$\frac{N}{\rho} - \frac{N}{\rho}e^{-\eta\frac{N}{\rho}} = \frac{N}{\rho}\eta \tag{2.91}$$

$$\therefore 1 - e^{-\eta\frac{N}{\rho}} = \eta \tag{2.92}$$

$$\therefore \frac{N}{\rho} = -\frac{\ln(1-\eta)}{\eta} \tag{2.93}$$

If we specify η we can therefore define:

$$z_+ = \eta\frac{N}{\rho} - \ln\left(\frac{N}{\rho}\right) \tag{2.94}$$

$$\therefore z_+ = -\ln(1-\eta) - \ln\left(-\frac{\ln(1-\eta)}{\eta}\right) \tag{2.95}$$

$$z_- = -\ln\left(\frac{N}{\rho}\right) \tag{2.96}$$

$$\therefore z_- = -\ln\left(-\frac{\ln(1-\eta)}{\eta}\right) \tag{2.97}$$

If we start with η we don't need to solve an iterative equation for $z_\pm$, so this method will significantly reduce computation time.

Also:

$$\frac{N}{\rho} = x_{\max} + 1 - z_- \tag{2.98}$$

$$\implies x_{\max} = \frac{N}{\rho} - 1 + z_- \tag{2.99}$$

$$\therefore x_{\max} = \frac{-\ln(1-\eta)}{\eta} - 1 - \ln\left(-\frac{\ln(1-\eta)}{\eta}\right) \tag{2.100}$$

So if we know $I_{\max}$ we can find:

$$\rho = \frac{I_{\max}}{x_{\max}} \tag{2.101}$$

2.2.9.3 *Eyam equation solver via and a GUI*

Let us summarise our Eyam equation solver using the $\eta = \frac{D_{tot}}{N}$ parameter:

$$z_+ = -\ln(1-\eta) - \ln\left(-\frac{\ln(1-\eta)}{\eta}\right) \tag{2.102}$$

$$z_- = -\ln\left(-\frac{\ln(1-\eta)}{\eta}\right) \tag{2.103}$$

$$x_{\max} = \frac{-\ln(1-\eta)}{\eta} - 1 - \ln\left(-\frac{\ln(1-\eta)}{\eta}\right) \tag{2.104}$$

$$\rho = \frac{I_{\max}}{x_{\max}} \tag{2.105}$$

$$\tau(z) = \int_0^z \frac{dz'}{x_{\max} + 1 - e^{-z'} - z'} \tag{2.106}$$

$$x = x_{\max} + 1 - e^{-z} - z \tag{2.107}$$

$$y = e^{-z} \tag{2.108}$$

$$t = \frac{\tau}{\alpha} + t_{\max}, \ \ I = \rho x, \ \ S = \rho y, \ \ D + R = \rho(z - z_-) \tag{2.109}$$

$$D = k(R+D), \ \ R = \frac{1-k}{k} D \tag{2.110}$$

$$N = I_{\max} + \rho - \rho z_- \tag{2.111}$$

$$R_0 = \frac{N}{\rho} \tag{2.112}$$

Although we know $0 \le \eta \le 1$, we still need a practical mechanism to determine what best fits a $\{t, I\}$ dataset. To achieve this we have developed the GUI concept described in the previous section relating to the Euler iterative method. A screenshot is provided in Fig. 2.5, with the Mompesson plague data as an example. Inputs of $t_{\max} = 0.900$, $I_{\max} = 26.76$, $\alpha = 2.987$ were set, and η was found to be $\eta = 0.6851$ when $\rho = 163.3$. As indicated in Fig. 2.5, the correct values of $I_0 = 14.5$ and $S_0 = 235$ are predicted, as is the total population of $N = 275$. Note the $D(t)$ curve has been shifted by 25 such that $D(0) = 0$, to enable comparison with the Mompesson data where the dead count also starts at zero when time is zero.

2.2.9.4 *Sum of squared differences*

To offer a quantitative measure of curve fit beyond visual correlation a 'SSD' is computed. This is the sum of the squares of the difference of model $I(t)$ values, interpolated at $\{t, I\}$ data times, to the corresponding I data values. In the above example SSD = 39.075. Alternatively, SSD can be the sum of squared differences between model predictions and all their data counterparts, i.e. if S & D exist, in addition to I, as independent measurements as in the Mompresson 1666 plague dataset. The Use S,D? GUI checkbox allows for both possibilities.

2.2.9.5 *Optimising $t_{\max}, I_{\max}, \alpha, \eta$*

The SSD metric can be used as the basis of an automated optimisation of Eyam model parameters. The principle is to create a *surface* based upon two of the four Eyam model input parameters $t_{\max}, I_{\max}, \alpha, \eta$, and determine the parameter pair coordinate associated with the minimum of this surface. An ideal situation would be to explore all four parameters in some form of four-dimensional space, but this would be somewhat difficult to visualize!

Which pair of inputs from should be chosen? The natural grouping appears to be $t_{\max}, I_{\max}$, i.e. visual characteristics of the $I(t)$ curve peak, and α, η, which relate to the width and skew of the curve. The Eyam solver GUI has two `Optimize` buttons which construct a mesh of parameters values, based upon the limits set by the sliders, and run the model over the full range.[19] Experience with the Mompesson, and also the 2014 Liberia Ebola data, suggests that the SSD surface based upon variation of α, η is problematic as an automated system, since the surface can be somewhat like a 'crumpled piece of paper' with quite a large range of what, by visual inspection using the GUI sliders, is the likely optimum. The well-defined peak of most $I(t)$ curves, even if the data is sparse and noisy, implies that the $\mathrm{SSD}(t_{\max}, I_{\max})$ surface is much more likely to have a clear minimum. This is certainly evident in Fig. 2.6(c).

2.3 Applying the Eyam Equations to Ebola

The S, I, R, D model outlined in the previous section was developed as an educational project, with the goal of incentivising students to engage in independent research and apply their mathematical skills to an interesting, and gruesomely contextualised, scenario. The model is, in a *Goldilocks & the Three Bears* porridge sense, 'just right'. There are enough variables to provide representational challenges, and the nonlinear nature of the Eyam equations provides a technical test. However, the system is by no means impenetrable, and indeed a significant portion of a pre-University mathematics course is applied in the analysis of the previous section, and will be applied in the stochastic model of the following section. However, following the sabbatical work of my colleague Dr John Cullerne (who began this Epidemiology of Eyam enterprise in the first place) at the Mathematical Institute of Oxford University, it might be possible that the semi-analytic model involving the $t_{\max}, I_{\max}, \alpha, \eta$ parameter quartet could be of more widespread utility in the field of epidemiology itself. A key test case is to see what happens if the Eyam model, with suitably modified inputs, could be used to characterise a modern epidemic. In *Extending the Epidemiology of Eyam* [5], we contrast the Mompesson 1666 plague epidemic with Ebola in Liberia in 2014. Although this is a very different disease, the $I(t)$ curves presented by the WHO Ebola response team nonetheless appear to follow the same 'slightly askew' singly peaked time variation as the Mompesson data, at least for the time bounds of a given outbreak.

Applying the methodology of the previous section in a MATLAB GUI (see Fig. 2.6), and following the optimisation process described above, the following

---

[19]In the Eyam GUI, a 20×20 parameter mesh is used.

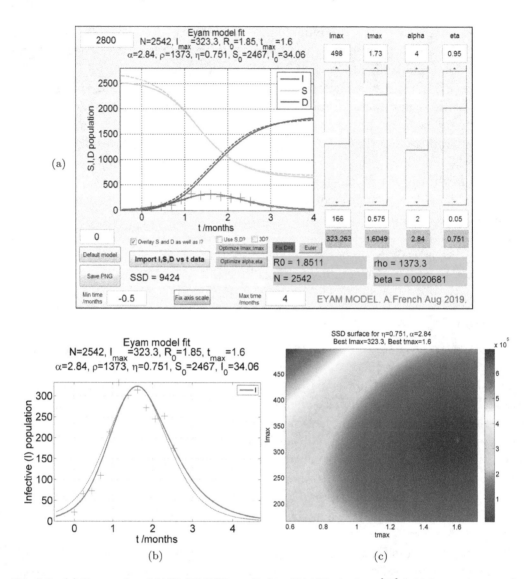

Fig. 2.6 (a) Eyam solver MATLAB GUI, applied to 2014 Ebola data [59], which constitutes only I vs t. The solid lines represent an Eyam model fit, and the dashed lines the corresponding K&K approximation. The fitting process was achieved via a combination of visual fit, using the sliders, and automatic optimization via varying a pair of inputs and working out the coordinate pair at the minimum of a surface formed from the Sum of Square Differences (SSD) between model and data I as a function of the inputs. (b) Eyam model fit to 2014 Liberia Ebola $I(t)$ data. Estimation of susceptible population N and Basic Reproduction Number R_0 from a single set of $I(t)$ data is thought to be rather useful in Epidemiology. (c) Optimal $\{t_{max}, I_{max}\}$ parameters, corresponding to the minimum of an SSD surface. α, η parameters were fixed in this automatic optimization. In the GUI (a), this is achieved by clicking on the **Optimize Imax, tmax** button. An equivalent is also available for α, η, which fixes t_{max}, I_{max}.

model input parameters were determined for the Liberia 2014 Ebola outbreak:

$$t_{\max} = 1.605, \ I_{\max} = 323.3, \ \alpha = 2.84, \ \eta = 0.751 \qquad (2.113)$$

and derived parameters:

$$\rho = 1,373, \ N = 2,542, \ R_0 = 1.85 \qquad (2.114)$$

Note this suggests that out of an 'at risk' population of $N = 2,542$, about 75% might ultimately die or recover from Ebola.[20]

For a suitably localised, and time bounded infection, the constant population Eyam model may well prove to be a reasonable model, and certainly one to try out as a sanity check before a more sophisticated multi-stage and multi-infection vector, or black-box artificial intelligence (AI) computation is attempted, models which might be much harder to verify given the practical difficulties in obtaining statistically significant quantities of quality field data. The big advantage of the Eyam model is its relative simplicity, and the ability to calculate important characteristic parameters with fairly trivial computational resources. Although something like our MATLAB GUI in Fig. 2.6 is recommended as a software system, it is possible to do it with a spreadsheet or, if you are particularly heroic, a standard-issue calculator.[21] If an $I(t)$ curve can be fitted to a dataset, you can not only predict $S(t)$ and $D(t)$, but you can also work out α, ρ, N, R_0. In the WHO Ebola Response Team report for the Liberia 2014 outbreak [59], $R_0 = 1.83 \pm 0.11$. It is encouraging that our prediction of $R_0 = 1.85$ is very much within this range.

2.4 The Importance of Basic Reproduction Number R_0

R_0 can be thought of as the *number of susceptibles converted to infectives, for every one infective, per unit of time* $\frac{1}{\alpha}$. You can see this from the 'Eulerisation' of the Eyam equation $dS/dt = -\beta S I$:

$$\Delta S \approx -\beta S I \Delta t \qquad (2.115)$$

$$\therefore \Delta S \approx -\beta S \times 1 \times \frac{1}{\alpha} = -\frac{S}{\rho} \approx -\frac{N}{\rho} = -R_0 \qquad (2.116)$$

So for $R_0 = 1.85$, this means Ebola will cause slightly less than two susceptibles to becomes infected for every infective, per unit time $\frac{1}{\alpha}$, which for our Ebola analysis is $\frac{1}{2.84} = 0.35$ months or ≈ 10.7 days.

R_0 is also directly related to a very important quantity, the minimum fraction $F_{\min}$ of the population to be immunized in order for 'herd immunity' (essentially

---

[20]In [5] we do not speculate on the mortality fraction k for Ebola. Taken literally, the D values in the paper correspond to a worst case scenario of $R = 0$ i.e. no possibility of recovery.

[21]The main problem for a spreadsheet or calculator is the numeric evaluation of $\tau(z)$. This is considerably easier to implement using a programming language.

a lack of susceptibles to catalyse an epidemic) to prevent the likelihood of a major epidemic:[22]

$$F_{\min} \approx 1 - \frac{1}{R_0} \tag{2.117}$$

We can justify this result with reference to the Eyam equations. For an infection to spread, I must increase with time. We showed in the previous section that for this to occur, the susceptible population $S > \rho$. Now assume at the start of an infection the total population $N \approx S_0$, i.e. $I \ll S$ and, we can also ignore any dead D and recovered R. Therefore for the infection to spread: $N > \rho$ or, since $R_0 = \frac{N}{\rho}$:

$$R_0 > 1 \tag{2.118}$$

The interpretation that R_0 represents the *susceptibles converted to infectives, per unit infective agent, per time unit*, implies that the probability of *no infection* must *decrease* as R_0 increases. If R_0 is large, this implies a highly contagious disease, which therefore ought to spread more rapidly and make an epidemic more likely. A simple mathematical rule which agrees with this statement is:

$$P(\text{epidemic doesn't develop}) = \frac{1}{R_0} \tag{2.119}$$

Now since $P(\text{epidemic spreads}) + P(\text{epidemic doesn't develop}) = 1$, this means:

$$P(\text{epidemic spreads}) = 1 - \frac{1}{R_0} \tag{2.120}$$

It makes relatively good sense that $F_{\min} = P(\text{epidemic spreads})$. If you have a high probability of an epidemic spreading, i.e. close to unity, then one would expect a much greater fraction of the population needs to be vaccinated to establish herd immunity.

For our Ebola modelling, $F_{\min} = 1 - \frac{1}{1.85} = 45.9\%$. For the Eyam plague (see Fig. 2.5), $R_0 = 1.68 \implies F_{\min} = 40.5\%$. It is instructive to compare Ebola and Plague to Measles. In *The Lancet* (2017),[23] R_0 for Measles is between 12 and 18. Using the upper limit, this means $F_{\max} = 1 - \frac{1}{18} = 94.4\%$. Such a high value highlights the importance of comprehensive population immunisation using the Measles–Mumps–Rubella (MMR) vaccine.

A slightly more rigorous argument can be used to justify $P(\text{epidemic doesn't}$ develop$) = \frac{1}{R_0}$. Define:

$$q_I = P(\text{epidemic doesn't develop from } I \text{ infectives}) \tag{2.121}$$

---

[22]https://www.healthknowledge.org.uk/public-health-textbook/research-methods/1a-epidemiology/epidemic-theory (accessed 27/8/2019).

[23]R_0 *and Measles*, reviewed in *The Lancet*. https://www.thelancet.com/journals/laninf/article/PIIS1473-3099(17)30307-9/fulltext (accessed 28/8/2019).

We can then write:

$$q_1 = P(\text{dead, or recovered})q_0 + P(\text{infection})q_2 \qquad (2.122)$$

i.e. in a time interval Δt, the probability of no epidemic developing from $I = 1$ infectives is the sum of the probability that the infective dies/recovers (i.e. 'removed' from the active population) and the sum that an extra infective is claimed from the susceptible population, thereby resulting in $I = 2$ infectives, and this pair does not result in the development of the epidemic. We need both parts since the Eyam equations state that in a time interval Δt, flows from susceptible to infective, and infective to dead (or removed), can occur simultaneously. We'll ignore contributions from multiple infectives claimed from the susceptibles, assuming this is an order of magnitude less likely.

Now $q_0 = 1$ since there are no infectives, and the Eyam model doesn't permit flow from D to I. Now if we accept the approximation:

$$q_2 \approx q_1^2 \qquad (2.123)$$

$$\implies q_1 \approx P(\text{dead, or recovered}) + P(\text{infection})q_1^2 \qquad (2.124)$$

We can use the Eyam equations to determine expressions for $P(\text{dead, or recovered})$ and $P(\text{infection})$. In time Δt the total change of population is $\beta S I \Delta t$ from the susceptibles to infective, and $\alpha I \Delta t$ from infective to dead. Therefore it makes sense to write the probability of infection to be the fraction of this total change which corresponds to a gain in the infective population, i.e:

$$P(\text{infection}) = \frac{\beta S I \Delta t}{\beta S I \Delta t + \alpha I \Delta t} = \frac{\frac{\beta}{\alpha}S}{\frac{\beta}{\alpha}S + 1} \approx \frac{N/\rho}{N/\rho + 1} = \frac{R_0}{R_0 + 1} \qquad (2.125)$$

Similarly:

$$P(\text{dead, or recovered}) = \frac{\alpha I \Delta t}{\beta S I \Delta t + \alpha I \Delta t} = \frac{1}{\frac{\beta}{\alpha}S + 1} \approx \frac{1}{N/\rho + 1} = \frac{1}{R_0 + 1} \qquad (2.126)$$

Note:

$$P(\text{infection}) + P(\text{dead, or recovered}) = \frac{R_0}{R_0 + 1} + \frac{1}{R_0 + 1} = 1 \qquad (2.127)$$

which makes good sense.

Hence:

$$q_1 = P(\text{dead, or recovered}) + P(\text{infection})q_1^2 \qquad (2.128)$$

$$q_1 = \frac{1}{R_0 + 1} + \frac{R_0}{R_0 + 1}q_1^2 \qquad (2.129)$$

$$\therefore q_1^2 - \left(1 + \frac{1}{R_0}\right) q_1 + \frac{1}{R_0} = 0 \tag{2.130}$$

$$\therefore (q_1 - 1) \left(q_1 - \frac{1}{R_0}\right) = 0 \tag{2.131}$$

$$\therefore q_1 = 1, \frac{1}{R_0} \tag{2.132}$$

We know from above that an epidemic can't develop if $R_0 \leq 1$. This scenario we can associate with $q_1 = 1$, i.e. there is total certainty that an epidemic will not develop from $I = 1$ infective. The second root of the quadratic for q_1 is for when the epidemic could develop (i.e. $R_0 > 1$), but due to good fortune, it doesn't. Hence if $R_0 > 1$:

$$P(\text{epidemic doesn't develop}) = \frac{1}{R_0} \tag{2.133}$$

2.5 A Stochastic Version of the Eyam Model

2.5.1 *Two problems with the Euler Eyam equations*

There is an obvious problem with the continuous S, I, D model[24] described in the previous section. We are dealing with human populations, which don't come in fractional quantities. In other words, the model above is an *idealization*, which one would assume would become more realistic if the total population size increases to the point that the change in S, I or D between each time step is to a very good approximation a whole number. However, the relative modest population of Eyam means this viewpoint is not so valid. We therefore need a *discrete* version of our model, i.e. where S, I, D populations are all positive *integers*. There is also an additional problem; transmission of plague (or any other disease) is likely to be a *random* process. In other words, any interaction between an infective and a susceptible may or may not result in disease transmission. However, we might anticipate the chance of transmission (i.e. the probability) to be governed by a rule which may take a similar form to the Eyam equations above. To make progress, we shall re-develop the Eyam equations in a similar, probabilistic, way that Brauer poses [4].

2.5.2 *A stochastic Eyam model as a Poisson process between time steps*

Assume the total population $N = S + I + D$ at time t and that this is a fixed quantity.[25] Assume timescales are short enough for natural birth and deaths to

---

[24] As in previous sections, D is actually $D + R$ (Dead + Recovered) population, for brevity.

[25] D could be shifted by a fixed quantity to make sure that $S + I + D$ is indeed the same N as defined in the previous section.

be negligible changes compared to the transition between infective I and dead D populations as a result of the action of the infection. Let us model the transmission of disease in a similar manner as one might model the exchange of energy and momentum between gas molecules. The probability of a random 'collision' between a single infective and the susceptible population is S/N. Hence the expected number of likely collisions given I infectives is $\frac{S}{N} \times I$. Now, as Brauer suggests, assume each interaction between infective and susceptible results in transmission of disease to fraction β of population N per unit interval of time. Hence expected change (i.e. a reduction) of susceptible population in time interval Δt is:

$$\Delta S = -\frac{SI}{N} \times \beta N \Delta t \tag{2.134}$$

$$\therefore \Delta S = -\beta S I \Delta t \tag{2.135}$$

Now let us further assume a constant fraction α of the infective population die per unit time interval. This means:

$$\Delta D = \alpha I \Delta t \tag{2.136}$$

Combining this loss of the infective population with the gain from the susceptible population:

$$\Delta I = \beta S I \Delta t - \alpha I \Delta t \tag{2.137}$$

We have arrived at a discrete version of the same Eyam differential equations used previously, and indeed when we employed the Euler method between discrete timesteps Δt, this version is actually identical in form. However, the difference now is that S, I, D are *strictly positive integers,* and the changes above are merely the *expected values* of a *random* process. As in the continuous, *deterministic* models,[26] the resulting infective curve should have a peak when the rate of flow of susceptible to infective populations equals the outflow of infectives to Dead (+ Recovered) populations when $S = \rho = \frac{\alpha}{\beta}$.

The probability of x discrete events occurring within a defined time interval, given there is a known expected rate λ, can be modelled by the *Poisson distribution*.[27]

---

[26] The solutions of the Eyam equations in the previous section are *deterministic* because the curves are always the same, given fixed input parameters. In a *stochastic* curve, random chance is incorporated, implying a range of possible curves, and different curves each time the simulation is run.

[27] The *Poisson distribution* is actually a limiting case of a *Binomial distribution*, which describes the probability of r occurrences of a *binary outcome* (e.g. A or B) in N *independent trials*, with probabilities $P(A) = p, P(B) = 1-p$. The idea is to divide up a time interval into a *large* number N of independent trials, where probability p is *small*. However, the expected number of A outcomes, $\lambda = Np$ is a modestly sized number. In a football example, the number of goals per match might be (based upon European Premier League averages) about 2.7. However there are very many kicks of the ball, and only a very small number of these result in a goal.

For example, the number of goals in a football match, lightning strikes per minute or yawns per hour during a particularly dry statistics lesson. If x is Poisson distributed:

$$x \sim \text{Po}\,(\lambda) \tag{2.138}$$

$$P(x) = \frac{\lambda^x}{x!}e^{-\lambda} \tag{2.139}$$

$$E[x] = \lambda \tag{2.140}$$

$$V[x] = \lambda \tag{2.141}$$

$E[x]$ is the expected value of x and $V[x]$ is the variance (i.e. a measure of deviation from the mean).[28] These quantities are the same for the Poisson distribution. The Poisson distribution of probabilities $P(x)$ is small for both small and large x values, peaking at $x = \lambda$. If λ is small, the shape of the curve is asymmetric (i.e. it has a *skew*), but tends towards a *Gaussian* bell-shaped curve as $\lambda \longrightarrow \infty$. In Fig. 2.7 (b), the probabilities $P(x)$ are compared to a normalised histogram of samples from the distribution. There is unsurprisingly a very good agreement! We can therefore incorporate 'just the right character of randomness' into our Eyam model by using the expected values of ΔS, ΔI and ΔD to define Poisson distributions for the discrete changes between time steps Δt.

$$\Delta S = -x, \ \Delta I_1 = x \tag{2.142}$$

$$x \sim \text{Po}\,(\beta SI \Delta t) \tag{2.143}$$

$$\Delta D = y, \ \Delta I_2 = -y \tag{2.144}$$

$$y \sim \text{Po}\,(\alpha I \Delta t) \tag{2.145}$$

$$\Delta I = \Delta I_1 + \Delta I_2 \tag{2.146}$$

The computational problem remains *how to sample* from a Poisson distribution. Although many computer-based statistics packages will offer this as standard, it is in keeping with the spirit of *Science by Simulation* to show how this might be achieved, and indeed a generic method for any discrete probability distribution.

2.5.2.1 *Generating random integers from discrete probability distributions*

The fact that the sum of the probabilities in a discrete distribution must sum to unity can be used to generate random integers from that distribution, assuming it is possible to generate a random number within the range [0,1]. (In MATLAB this is what the **rand** function generates). The idea is to use the probabilities to form the edges of a series of 'boxes' which, when joined, span the interval [0,1]. For every random fraction $\sim \text{U}(0,1)$, determine the box number which encloses the

---

[28]The *expectation* (mean) of discrete values of quantity x_i is: $E[x] = \sum_i x_i P(x_i)$. The *variance* is: $V[x] = \sum_i x_i^2 P(x_i) - (E[x])^2$.

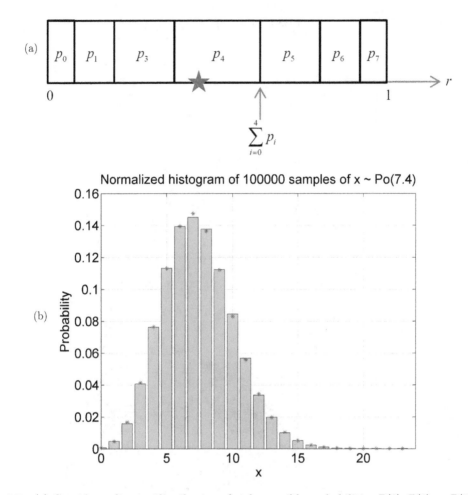

Fig. 2.7 (a) Consider a discrete distribution of eight possible probabilities $P(0), P(1)\dots P(7)$, for the random integers $0,1,\dots,7$. The widths of eight joined boxes correspond to these probabilities. A random number $r \sim \mathrm{U}(0,1)$ is chosen. In the diagram this happens to be in the range $\sum_{i=0}^{3} P(3) \leq r < \sum_{i=0}^{4} P(4)$, so in this case the random number $x = 4$ is generated. (b) Example of the Poisson probability distribution. The red stars represent the probabilities $P(x) = \frac{7.4^x}{x!} e^{-7.4}$ whereas the green bars describe the normalised histogram of 100,000 random samples from the distribution $x \sim \mathrm{Po}(7.4)$. *Normalised* means the total frequency (i.e. area of the histogram) sums to unity.

fraction. This box number is the random variable x, and the process is illustrated in Fig. 2.7(a).

2.5.2.2 *Combining the Euler and Poisson Eyam models*

The ideas above have been implemented as a MATLAB function to enable maximum automation and computational efficiency. Ideally the α, β parameters should be determined via some form of 'sum of squared differences' method, separately for Euler and Poisson techniques. However, to save time I have decreed that the Euler

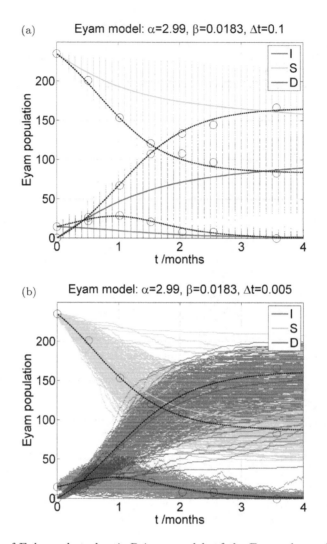

Fig. 2.8   Overlay of Euler and stochastic Poisson models of the Eyam plague. In all cases inputs are: $I_0 = 14.5$, $S_0 = 235$, $\alpha = 2.99$, $\beta = 0.0183$. The black circles represent the Mompesson data, the dashed black lines represent the Euler result and the dots represent 200 overlaid iterations of the Poisson model. The red, green and blue solid lines represents a mean average of the 200 Poisson iterations. (a) In this case $\Delta t = 0.1$ months, and the Poisson trends are somewhat different to the Euler predictions. Indeed the average Poisson infective line has no peak. (b) In this case $\Delta t = 0.005$ months, and the average Poisson trends are similar to the Euler predictions, although they don't track exactly the same curves.

model be used to determine the optimal parameters, given the timestep Δt. The Poisson model is then run using the same values and thus a fair comparison can be made. All other Eyam populations parameters are the same as described in the previous section. In Fig. 2.8(a) the timestep $\Delta t = 0.1$ months, and there is little agreement between the Euler and mean average of the overlay of 200 runs of the stochastic model. However, in Fig. 2.8(b) the timestep $\Delta t = 0.005$ months and the

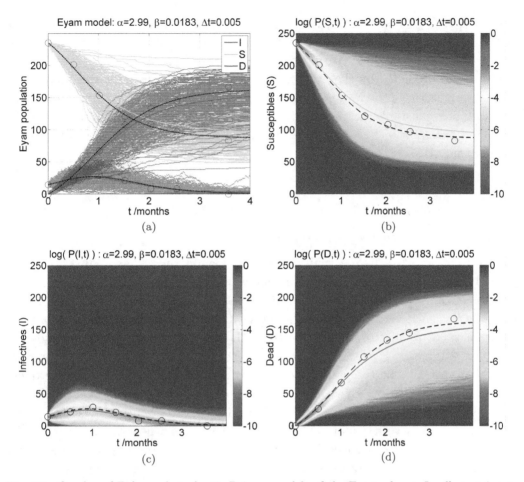

Fig. 2.9 Overlay of Euler and stochastic Poisson models of the Eyam plague. In all cases inputs are: $I_0 = 14.5$, $S_0 = 235$, $\alpha = 2.99$, $\beta = 0.0183$. The black circles represent the Mompesson data and the dashed black lines represent the Euler solution, using $\Delta t = 0.005$ months. (a) is an overlay of 200 stochastic traces, with the solid lines being the average. (b)–(d) represent probability maps, determined from running the stochastic model for 50,000 iterations. At each timestep, a histogram of population values for S, I, D is determined. This is then normalised to yield a set of probabilities for each population value. The logarithm of these probabilities is used to determine the colour of the figure, and hence enable the probability of S, I, D populations vs time to be visualised. As in Fig. 2.8, there is an small difference between the average and Euler solver solutions.

curves appear to be much closer. One might sensibly expect the curves to align as Δt reduces further, although in Fig. 2.9, which compares the 200 iteration result (a) to 50,000 iterations (b,c,d), it appears that the mean average of the stochastic curves are slightly, but quantifiably different to the 'semi-analytic' or Euler method solutions of the previous sections. At the time of writing, I am still not sure of the reason for this. The fact that the difference appears to be preserved as Δt shrinks indicates that it is perhaps more fundamental than a numerical method artefact.

In summary, the stochastic, discrete Poisson population model obeys the same essential equations as the continuous version (which is albeit evaluated in a fixed time-step iterative manner). We can therefore ascribe some confidence in the approximate reality predicted by the Euler model, and the mechanism for the inference of α, β parameters from the Mompesson data. However, the stochastic model affords us a much more realistic description of the Eyam population flows, and importantly, the predicted *variation* of S, I, D vs time given the random nature of the transmission of infection.

2.6 Applying the Eyam Equations to COVID-19

2.6.1 *The coronavirus pandemic in 2020*

Memories of late 2019 to 2021 will certainly be dominated by a global pandemic of a novel flu-like respiratory disease, a *coronavirus* designated COVID-19. It is named because of the seventy-four or so pin-shaped 20 nm spike-proteins that emerge from the spherical surface of the virus. These give its electron micrograph image a similar appearance to that of the *solar corona*, the aura of plasma that surrounds our sun that is typically only visible during an eclipse. The disease was first reported in Wuhan in China at the end of 2019, and by April 2020 cases were being documented across the globe. By January 2021, two million deaths had been attributed to COVID-19, and 90,000 in the UK. Many nations experienced 'lockdown' measures of varying severity, and this chapter was written when the UK entered its third period of enforced social distancing to help arrest the spread of the virus. Developing COVID countermeasures have understandably been a huge focus for the scientific community, and the formulation of multiple vaccines in matter of months rather than years is a triumph of human ingenuity and collective action. Following the vaccine development and production process, the baton was then passed to an army of vaccinators and logistics support teams who would deliver millions of doses over the following few months to the most vulnerable, and on a slightly longer timescale to those less at risk. Underpinning the whole enterprise was a global data collection effort which must surely exceed in scope anything previously attempted in epidemiology. Oxford University's 'Our World in Data' online platform https://ourworldindata.org/coronavirus [41] is a particularly comprehensive resource, and is updated daily. The increased free availability of quality scientific data is surely a genuine positive in this difficult period, and one hopes this is something that will retain the status of 'new normal' once the world is comprehensively vaccinated. A robust response to the problem of climate change from human activity, and combating the social media vectored pestilence of 'fake news' that has addled the minds of so many of us in recent years, can only be aided by readily accessible scientific data, or, as I like to call it, *quantitative truth*.

2.6.2 *Infographics should match the narrative*

On a daily basis, bookshelves of COVID-19 analysis are published, so what can we add here? I think there are a few aspects of COVID-19 reporting, particularly in the early phases of the pandemic, where a greater degree of clarity could have been achieved. Although graphs and infographics have dominated news stories for much of the past year, many of the more subtle messages have been confusing. I think one of the causes is ambiguity in, and omission of, important numerical details. Many graphs don't have axes scales, which can make like-with-like comparison difficult, and can lead to the confusion of 'signal with noise', as Nate Silver might put it [44]. If one takes a small sample of data and zooms in on the various undulations, it is difficult to assess the significance of the displayed variation without scale-reference to overall trends. Some curves have been used to pictorially justify statements like 'we must reduce the infection peak' when the presented curves themselves, typically a scaling of the cumulative deaths $D(t)$, don't actually have a peak but instead follow characteristic S-shapes between plateaus. Using the Eyam equations as a basic model, we know the infection peak of $I(t)$ corresponds to the steepest *gradient* of $D(t)$ since $I \propto \frac{dD}{dt}$. If the picture (relating to $D(t)$) doesn't match the associated narrative (relating to $I(t)$), then it is unsurprising that less meaning is conveyed, and the public health message is potentially diluted in the resulting ambiguity. There is also the all-pervasive R_0 number, which appears to have a somewhat different meaning to what was defined in earlier sections of this chapter. Pandit [32] discusses many different possible incarnations, and I think the narrative from most UK COVID-19 official briefings matches a definition of:

$$R_0 = \frac{\delta I}{I} + 1 \tag{2.147}$$

where δI is the fractional change (per day) in estimated infectives I. This means $R_0 > 1$ implies a rising I, and $R_0 < 1$ means a reducing I. This is clearly *not* the same as $R_0 = \frac{N}{\rho}$, which applies to a *closed system* of total population N, affected by an epidemic with rate parameters α, β. The latter implies a virus & population characteristic of a given $I(t)$ curve, rather than a measure of how the $I(t)$ curve is changing on any given day.

Combatting COVID-19 is a comprehensive social effort, and the very fact that the vast majority of citizens have accepted really quite draconian curbs to their liberty without excessive state intervention, is hopefully a testament to the power of altruism over self interest. However, this does mean that the social contract between state and citizens must be upheld, and I think a critical aspect of this is the *basing of policy on facts that can be independently verified*. Failure to achieve this leads to mistrust in government, which can lead to a lacklustre response to urgent public messages, belief in conspiracy theories, and in extreme cases, insurrection. It is clearly absurd to assert a causal connection between the anti-vaccination movement, or the storming of Capitol Hill in January 2021 by a spear-wielding self styled shaman of the QAnon cult, and a lack of precise numeric information in a few graphs published on a news website. However, one must not underestimate the power of an accumulation of small uncontested fictions into the social consciousness. They can be very difficult to debunk once they have taken root. The curse of the internet is that falsehoods can spread at the speed of light in an optical fibre, much faster than any microbiological agent. The counter to fake news culture, populism and self-confirmation bias begins by asking yourself two simple questions: "What *exactly* is being said here?" and "How do we *know* it is true?" But you must be prepared to say: "I'm not entirely sure" or "what is written is not what I expected it to be," and "hmmm, this sounds like fact-garnished opinion." To put it bluntly, we need to re-teach ourselves to see the world as it actually is, not what we imagine it to be, and use all our powers of critical reasoning to make sense of our observations.

2.6.3 *Can the Eyam equations model COVID-19?*

To investigate the possible application of the Eyam model, I have used data downloaded from https://ourworldindata.org/coronavirus [41], which covers the time period from 31 December 2019 till 5 August 2020. This corresponds to the 'first wave' of UK COVID-19 infections. I am writing this in January 2021 during the 'second wave'. Unlike the Ebola and Mompesson Plague datasets, the Oxford COVID-19 data records cumulative deaths over time, i.e. $D(t)$, attributed to COVID-19, and also a log of total cases, rather than Infectives $I(t)$. I shall assume 'total cases' means $I + R + D$, i.e. the sum of infective, recovered and dead sub-populations. Let us assume that the basic ideas represented by the Eyam equations can be broadly applied to COVID-19, in particular the concept that we can *estimate* I from the *time derivative* of $D(t)$. If we do this then we need to know *a priori* the *mortality fraction* k, and the time constant α. Modelling COVID-19 via the Eyam equations is certainly an enormous oversimplification, particularly since the assumption of an isolated fixed population $S + I + R + D$ is unlikely to be correct, even during a national lockdown. However, mindful of these very important caveats, let us apply a partial Eyam analysis and see whether any useful insights can be determined.

The first Eyam equation gives: $\frac{d}{dt}(R+D) = \alpha I$. Since $D = k(R+D)$ this means:

$$I = \frac{1}{k\alpha}\frac{dD}{dt} \tag{2.148}$$

$$R = \frac{1-k}{k}D \tag{2.149}$$

To minimise noisy data spikes that result from the daily update, rather than continuous time history of $D(t)$, the time derivative $\frac{dD}{dt}$ shall be evaluated from a 7-day mean average. The output of this average corresponds to the sum of the reported death totals three days before, three days after and also the day which the

D value corresponds to, divided by seven. $\frac{dD}{dt}$ is approximated from the (daily) data using the numerical method:

$$\left. \frac{dD}{dt} \right|_{t_n} \approx \frac{1}{2} \frac{D_{n+1} - D_{n-1}}{(t_{n+1} - t_{n-1})} \qquad (2.150)$$

Throughout Oxford's 'Our World in Data' is the strong health-warning *not* to assume that the mortality fraction k is equal to the D values divided by the total cases. In most countries, and this was unfortunately particularly true for the UK during the first wave in 2020, testing for COVID-19 has been far from comprehensive, and therefore the number of recorded cases is possibly a significant underestimate of the true value of $I + R + D$. To run the Eyam model, and estimate I and R, I have taken estimates of k and α based upon research published in *The Lancet* [56] and *Nature Medicine* [61]. I shall set $k = 0.014$, i.e. 1.4% is taken as an average mortality fraction for COVID-19. Both Verity *et al* and Wu *et al* allude to significant variations across ages, ethnic groups and degrees of prior health condition, so one must use these figures with caution. The average time from infection till recovery or death is estimated to be 17.8 days. This means $\frac{1}{\alpha} = \frac{17.8}{365/12}$ months, i.e. $\alpha = 1.71$ months^{-1}. In Fig. 2.10(a), the deaths (in millions) are contrasted for nine countries. As of 5/8/2020, all appear to have reached a plateau of the S-shaped $D(t)$ curve, with the exception of the USA and Brazil. The PWD figure of 36.01 million is the predicted estimate of total world deaths (from all causes) over the period 31/12/19 to 5/8/20. This is based upon the estimate of 7.7 per 1000 per year,[29] and a total world population of 7,794,798,729. All populations are those stated in the Oxford dataset (which are fixed over the time period considered). Although the personal tragedies of COVID-19 cannot be overstated, it is nonetheless worth pointing out that the total cumulative world COVID-19 deaths per capita (at about 0.01% as of August 2020) themselves represented about 1 in 46 of total deaths[30] during the same period. COVID-19 is sadly one of many causes of death, and in our desire to mitigate the disease, we must not forget heart disease, malnutrition, cancer and other extant life-limiting mechanisms.

The estimated infectives divided by population for the nine chosen countries, and the world overall, is plotted in Fig. 2.10(c). This is perhaps the most revealing curve, as it allows the response to the virus by different countries to be assessed more fairly, factoring in significant variations in population. Italy and the UK have broadly similar curves, with a delay of peak infection of just under a month. France bisects

---

[29] https://en.wikipedia.org/wiki/Mortality_rate.

[30] To place this figure in context of more recent events, BBC News reported on 12/01/2021 that a total of 697,000 UK deaths in 2020 represented 85,000 more than the expected value based upon an average of the past five years. If all of these 'excess deaths' were attributed to COVID-19, this represents $\frac{85}{697-85} = \frac{5}{36}$ or about 14%. Note $\frac{1}{46}$ is about 2.2%, so it is sadly likely that COVID-19 may well prove to be far more deadly than this initial estimate might suggest.

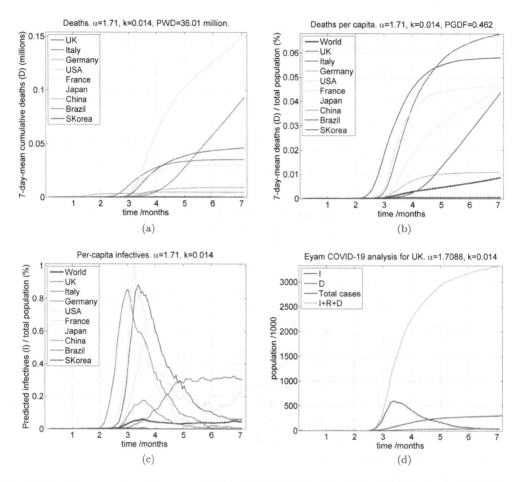

Fig. 2.10   Eyam equation ideas are applied to COVID-19 epidemiological data from Oxford University's 'Our World in Data' platform [41]. Deaths D (a) and deaths per capita (b) are 7-day mean averages of the cumulative deaths, reported daily for each country over the period 31/12/19 to 5/8/20. Infectives (c) are *estimated* from D based upon $\frac{d}{dt}(R+D) = \alpha I$ and $D = k(R+D)$. α and k parameters are estimated based upon research published in *The Lancet* [56] and *Nature Medicine* [61]. (d) is the Eyam model for the UK, and suggests that the total recorded cases (equivalent to $I + R + D$) is likely to be a significant underestimate. Note this suggests a calculation of mortality fraction k based upon D divided by total cases is also likely to be an overestimate.

these, and has a slightly higher infection peak, but also a reduced timescale (of just over two months) for the bulk of the infection. Germany has a much lower per-capita infection, and China, Japan and South Korea are barely visible on this graph. Brazil appears to be anomalous, with a plateauing infection, and the USA appears to be entering a second wave of infection.[31] I will leave a detailed commentary of

---

[31]Unlike most northern European countries, the USA covers a wide range of latitudes. The 'double-hump' is perhaps indicative of the $I(t)$ curve being a *superposition* of an initial Eyam-style curve for the Northern states, followed by a lower, and wider duration curve, for the Southern states.

why the per-capita infection curves are the way they are to the epidemiologists, politicians and professional pundits. However, it is clear from the data that the spread of the first wave of the virus in Germany, and certainly in Japan, China and South Korea, was far less than in the UK, Italy, France and the USA. There are some very important caveats that are worth repeating:

- The Eyam model assumes a fixed population $N = S+I+R+D$. This might not be such a bad model for a country under strict lockdown, but is unlikely to be true in the early phases of the pandemic; before borders were closed and social distancing measures, mask wearing and extra hygiene protocols were broadly adopted in most countries.
- I am assuming the Oxford dataset, particularly the count of cumulative COVID-19 deaths, is accurate for each country. As stated previously, it is likely that the 'total cases' is a significant underestimate, so my I estimates are based solely upon $\frac{dD}{dt}$. However, it is possible that the deaths attributed to COVID-19 may be erroneous, and indeed various corrections were applied during 2020. Most COVID-19 deaths appear to be associated with those with 'underlying health conditions' and therefore COVID-19 will not be the only factor in mortality. Establishing COVID-19 as a clear cause of death, and then working out what to contribute to population-averaged national statistics, is certainly a very difficult task. However, the very fact that the Eyam model predictions for $I(t)$ yield sensible numbers, can perhaps be used to boost confidence in the dataset. Either way, the myriad in-built uncertainties must be taken into account, and ideally these should be represented on the $I(t)$ graph too. This being said, a rigorous data-driven model with (bounded) uncertainties is far better than no model at all, and certainly better than any decision making based upon non-scientific (anti) reasoning.

Figure 2.11 illustrates an Eyam Equation analysis of the I, R, D estimates from the Oxford COVID-19 data for the UK. It is very clear that the I curve is a poor fit, and that the curve has a much steeper gradient prior to the infection peak than any η and α parameter combination will predict. A multitude of factors could explain this difference. As stated above, the assumption of a constant total population is unlikely to be true in the early phases, i.e. before comprehensive lockdown was introduced at the end of March 2020. Also, the progressive introduction (and then easing) of lockdown measures may have 'widened' the $I(t)$ curve after the peak. Perhaps a *different* η parameter should be applied before and after. In Figs. 2.11(a) and 2.11(c) I have used $\eta = 0.774$ to enable an asymptotic agreement of the R curves with the estimates from the Oxford data. Note R in the model is actually $R + D$, but given $k = 0.014$, we can neglect D for the purposes of approximate curve-fitting. In this scenario $N = 4,254,000$ and $R_0 = 1.92$. In Figs. 2.11(b) and 2.11(d) I have modified η to be 0.912 to enable a better fit of the post-peak I curve.

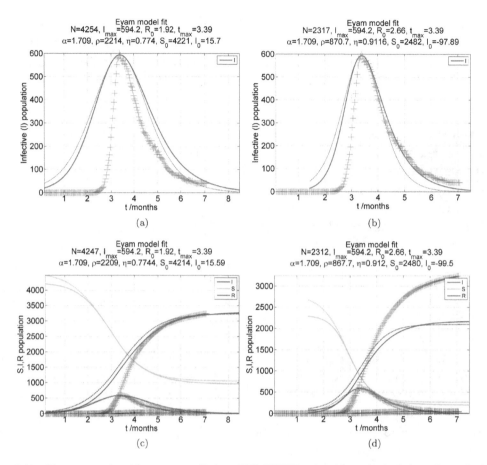

Fig. 2.11   Eyam equation ideas are applied to UK COVID-19 epidemiological data from Oxford University's 'Our World in Data' platform over the period 31/12/19 to 05/08/20 [41]. Populations are in 1,000s. The infective curve, estimated from a numeric evaluation of $\frac{dD}{dt}$, is poorly fitted by the Eyam model. Data are + symbols, models are represented by curves. The dashed curves are the K&K approximation. The fit in (a) and (c) is based upon η set such that the R curves fit the data asymptotically, whereas (b) and (d) use an η value to fit the post-peak infective curve.

This yields $N = 2,312,000$ and $R_0 = 2.66$, but underestimates the $R + D$ population and predicts a negative I_0 value, which has no physical meaning.

2.6.4   *Why I is more useful than R_0*

In the mass media there is much discussion of reducing R_0 to below unity to prevent the spread of the virus. I think this is potentially confusing, as in our Eyam model, $R_0 = \frac{N}{\rho}$ is characteristic of a given fixed-population outbreak, and not something that varies during its course. A Susceptible population S that is less than threshold $\rho = \frac{\alpha}{\beta}$ means 'beyond the peak'. For R_0 to vary, this implies a different characteristic infection curve at any point in time. As described above, I think $R_0 = \frac{\delta I}{I} + 1$ is a better match to the recent narrative, although it doesn't make sense how this could be used to compute the herd immunity fraction $F_{\min} = 1 - \frac{1}{R_0}$. An R_0 value of less than unity would mean $F_{\min} < 0$, which clearly has no physical meaning. What is worrying to me is that there should be any ambiguity at all. If a single number is supposed to characterise a globally important phenomenon, then we really ought to agree on its definition and how it is calculated.

I think a far better metric is to use I vs time. It is clear from the curve what 'peak infection' means, and the time variation of the curve gives a reasonable description of the dynamics of the infection. Although the Eyam equation methodology presented here is crude, perhaps using $I = \frac{1}{k\alpha}\frac{dD}{dt}$ (per capita, as in Fig. 2.10(c), and based upon 7-day averaged $D(t)$) could be a clearer metric for public health information. Cumulative deaths, new deaths or total cases don't really match the narrative of peaks and curve flattening, which apply to I vs t. For the narrative to align with the statistics, *we must talk about the same quantity to avoid a confusing message*. It is a bit like presenting calculations involving energy or momentum, when you are actually talking about force or acceleration. They are clearly related, but are *not* the same quantities and therefore cannot be interchanged in a linguistic sense. This is why the language of mathematics is so important: it prevents us from making semantic errors resulting from a loose use of technical speech. In the Oxford dataset, a variety of graphs is presented. In my opinion, the one that has most meaning is the '7-day averaged daily death rate, per million'. This is of course a direct scaling of $\frac{dD}{dt}$. The difference between the Oxford graphs and the ones presented here is that I have used 'deaths per capita per month' rather than per day, and per capita rather than per million. In other words, the relative proportions between the country curves should be identical to those in Fig. 2.10, but the absolute values scaled by fixed ratios. As pointed out very clearly in 'Our World In Data', there is a real danger of misrepresenting the dynamics of an infection if one reports 'confirmed infectives' rather than something proportional to $\frac{dD}{dt}$, since comprehensive testing is unlikely to be widespread. As mass-testing increases, one may reveal more of the (previously unaccounted for) infectives, while the actual total is rapidly dropping. You could be deluded that another 'wave' of infection is coming, whereas this is unlikely to be the case if you look at $\frac{dD}{dt}$ and this is falling exponentially.[32]

2.6.5 *Summary*

A highly simplified understanding of the dynamics of a disease in a closed population, i.e. where the Eyam equations are deemed to be a representative model, is unlikely to offer professional epidemiologists many new tools to battle a pandemic, but it

---

[32]This was quite a common criticism in the UK around August to September 2020, when deaths per day were much lower than they had been at the peak of the first wave in March–April 2020, but confirmed infections were rising rapidly. However, a second COVID-19 wave *did* certainly occur in the UK, compounded by a more virulent mutation of the virus. The $I(t)$ curve from January 2020 to January 2021 (see the latest 7-day average daily death rate, per million from *Our World In Data*) has a very clearly defined second peak.

might provide a wider range of citizens with the knowledge and confidence to ask more targeted questions of clarification. As proposed above, the counter to fake news culture, populism and self-confirmation bias surely begins by asking yourself two simple questions: "What *exactly* is being said here?" and "how do we *know* it is true?" An understanding of the $I(t)$ curve and its relationship to $D(t)$ will help frame the first question, and knowledge of the assumptions and limitations of the Eyam model may help frame the second.

A pandemic is not just a biological problem; it affects society and culture too. The intervention of worldwide mass social distancing is possibly unprecedented in human history, and it remains to be seen what long-term impact this has. There are also many tough rationalist lessons from a pandemic. One might be to resist the temptation to immediately assert a correlation, or perhaps even a cause, between a deep personal tragedy such an awful disease, and national policy. Although we may have well founded ire, it may not always be the Government's fault, and a single data point, no matter how individually important, may not always form a trend. We are all citizens in a human population of billions, and perhaps we ought to offer greater respect to the collective over the self, particularly in Western societies. But we must never forget that *every* data point on a $D(t)$ curve represents a life lost, and a broken connectivity, via social bonds, to many other lives.

As in all challenging activities, a measured response depends upon a rational calculation of risk: i.e. the *impact* of the risk multiplied by the *probability of occurrence*. The *numbers matter*, and must be placed in the context of other risks. Opinion, feelings and impatience may need to be overcome before one can start to plan the optimum intervention. Although it is the obvious priority response in a time of international crisis, an entirely myopic focus upon COVID-19 may be at the cost of greater (per capita) increased deaths due to untreated heart conditions, cancers, missed operations, delayed investment in antibiotics and other anti-virals, and not least the long term impact of a fractured economy and social fabric. Governments must continue to run (and publish) the numbers, and be absolutely clear when presenting the evidence behind policy decisions, particularly when no option is universally positive for all stakeholders. One of the major side-challenges of COVID-19, and perhaps a silver-lining of opportunity too, is to accelerate the fixing of the other extant problems of our lives, so we have sufficient time and resources to deal with the pandemic. If one is able, I think this is a call to action, whatever our role in society. The problems of extreme social inequality, other diseases, and climate change are not going to disappear while the world battles coronavirus.

2.7 A Physical Representation of the Eyam Model

To augment the story-context, numerical and analytical aspects to the introductory calculus pedagogy, it is also informative to borrow an idea from Economics that provides a kinaesthetic lens. With diseases it is perhaps somewhat irresponsible to perform actual experiments! The Phillips Machine or Monetary National Income

Analogue Computer (MONIAC) [33, 37] is a hydrodynamic 'computer' that models an economy[33] using flows of water to represent the flows of money, and pulleys, levers and reservoirs to model the inner workings. In Fig. 2.2 on page 32 we show the Winchester College Phillips Machine variant to illustrate the Eyam plague outbreak. The reservoir at the top is for "Susceptibles"(S) and the rate of flow depends on the level of water, marked by a floating ping-pong ball. The beaker is the "Infective" (I) reservoir, which is actually on a lever and righted by a mass and pulley, so, under certain conditions, if the rate of flow out of I is greater than the inflow, the lever and pulley will shut off the supply from S. Dr John Cullerne, who built the system with Andrew Chesters at Winchester, found that this visual model was particularly effective at conveying the associated dynamical ideas for students who lacked prior calculus knowledge. The system, quite literally, demonstrates flows between reservoirs. The students could see the system working in real time, and the movement of levers, pulleys and water levels beautifully illustrated the phenomenon as a whole. At the time of writing, we are working with our students to improve the system, perhaps with electronically controlled valves that faithfully simulate the actual flow rates defined by the Eyam equations. In a purely mechanical system it is somewhat tricky to calibrate and vary α, β parameters in a systematic way.

2.8 Extensions to the Eyam Model to More Accurately Model the Black Death Pandemic

2.8.1 *Bubonic, pneumonic and septicaemic plague*

Our Eyam model fails to take into account that the 1666 plague was split into three, connected variants: *Bubonic*, *Pneumonic* and *Septicaemic* [63]. Each has varying death rates. The Bubonic plague could evolve into either of the other two forms, and the Septicaemic plague could eventually progress into Pneumonic plague. As such, the Infectives population should be split three ways. Additionally, it is known that the plague was, in general, spread by both vermin and fleas [68]. A second population can be devised, labelled the *vector population*, in which individual vectors could *infect each other* through contact. The number of infective vectors could directly influence the rate of change of human Infectives. This is a much more complex model than the two-parameter Eyam system presented previously, and possibly a more complete explanation of a plague outbreak. However, to determine all these parameters, significantly larger amounts of data would be required, and the broad features of the model would be much harder to intuit.

2.8.2 *Quaranta giorni*

During the events of a plague outbreak, the city-state of Ragusa (modern Dubrovnik) established a rule where newcomers to the city were to be confined to nearby islands

---

[33]Albeit highly idealised. See important criticisms by Raworth [7].

for 40 days (initially 30). The Italian translation for '40 days', *quaranta giorni*, would then become the origin of the word *quarantine* [27], used today to refer to the isolation of Infectives to prevent the spread of a disease during an epidemic. To incorporate *quaranta giorni*, Susceptibles are introduced via the 'Quarantined Susceptibles' population, in which they must stay for a time of 40 days. One could vary the rate of people coming into the city, or indeed the quarantine period, to see how these additional variable changes affects the S, I, R, D populations. An even more accurate system might be to abandon the continuity of differential equations and use probability models of discrete 'agents' that represent the actual individuals, as suggested in the 'Stochastic Eyam' model variant. The S, I, R, D quantities are then sums of the agent population labels. The idea is to track an individual agent's progress through simulation time, rolling various (electronic) dice at each iteration to determine whether a population group change occurs (e.g. from S to I). If an agent is in quarantine, then he/she cannot contribute to the probability of a non-quarantined agent changing from S to I. In a similar manner to the Eyam model, we might anticipate the probability of a S to I change to be an increasing function of the product of S and I. However, the requirement that probability is in the range [0,1] means a simple linear relationship will be inappropriate in this case. Perhaps some form of S-shaped *sigmoid* function (with fairly linear gradient between extremes) might be a sensible candidate.

2.8.3 *Variolation, vaccination and modern medicine*

The Black Death occurred during the mid-fourteenth century. Certain preventive measures (i.e. use of antibacterial cleaning, vaccination), were not available until the eighteenth century, when Lady Mary Wortley Montagu brought *variolation* to Britain [65], followed by the safer *vaccination* methods developed by Edward Jenner and others [64]. If a Black Death outbreak were to occur in our modern setting, it is assumed that vaccination would be widely available for everyone. However it is naive to assume all human populations enjoy the same level of access to effective medicine. Also, could even the wealthiest and most scientifically enlightened countries respond quickly enough to outbreaks of previously unknown but highly contagious pathogens, to avoid a pandemic? Recent experience of COVID-19 would suggest otherwise.

One of the goals of epidemiology is to model the potential impact of limited funding and access to medicine. In particular, the human health impact versus cost of mass vaccinations. Is it necessary to vaccinate all, or can pandemics be prevented by shielding a carefully targeted subset of the potential Susceptible population? Note in many rural scenarios one might assume rate of flow from the Susceptible population tends to a (non zero) constant, as only a certain number of people can be vaccinated in a given day due to manpower constraints. It may not be feasible to completely eliminate a pathogen, although it is surely a very laudable aim to try.

Chapter 3

Holmes and Watson Meet Bayes

3.1 The Calculus of Chance

Almost every aspect of our lives is influenced by the 'calculus of chance'. In other words, most decisions that affect our future are based upon a calculation of the *probability* that a particular outcome will occur, *given* our current knowledge or set of assumptions. Probabilities are inherently *conditional*, even apparently obvious ones. "What is the probability that I roll a six with a six-sided dice, *given that it is fair*?" "What is the probability that an electron will 'tunnel' out of a semi-conductor in a computer chip, *given* our knowledge of the design and physical characteristics of the chip?" The inference here is that the 'givens' are essentially axiomatic, i.e. taken as 100% truths. In many practical scenarios, these givens may be themselves subject to significant uncertainty: "What is the probability that a Typhoon will head towards Hong Kong tomorrow, given what I know about its current and historical dynamics?" "What is the probability that Moriarty is guilty of multiple crimes most foul, given the circumstantial evidence determined by Holmes and Watson?"

The mathematics of *conditional probability*, particularly the ideas perhaps first articulated by the English statistician, philosopher and Presbyterian minister Thomas Bayes (1701–1761), are therefore some of the most useful in our toolset, and certainly fundamental to how we articulate a quantitative prediction about the future. However, books like *The Signal and the Noise* [44] and *Irrationality* [51] suggest that most of us, and even some professionals, perpetuate a misunderstanding of the core ideas. This can lead to undesirable consequences, from misguided economic policies to miscarriages of justice, to erroneous assessment of suspected disease. We shall explore the concept of *Bayesian inference* using the latter as context.

3.2 Probability, Hypotheses and Tests

Before we work through an example, let us define some some general notation associated with conditional probability. Let H be some hypotheses that can be tested in a binary true/false fashion via some test. $P(H)$ means the probability that H is true and $P(H')$ the probability that H is false. $P(H|T)$ is the probability that H is true *given* a positive test outcome T i.e. a 'pass.' Conversely, $P(H'|T')$ is the probability that H is false, given a test fail T'. There are also two other possibilities: $P(H|T')$ is the *false negative*, i.e. the probability H is true given a test fail T'. $P(H'|T)$ is the *false positive*, i.e. the probability H is false given a positive test T.

3.3 Diseases, Tests and Conditional Probability

A worried male patient enters a doctors surgery with flu-like symptoms. After significant internet research, the patient convinces himself that he may have a rare brain degenerative disease, which apparently manifests itself in 1% of patients with similar symptoms. He also reads that there is a nose-fluid test for the disease which is claimed to be 95% accurate. The patient begs the triage nurse to run the test, and to his horror, it comes back positive! Should he despair? Well possibly, given anecdotal evidence of acute hypochondria, but the truth of the matter is the probability of him actually having the disease, *given the test being positive* is actually only about 16%. The key point is the answer is *not* 95%, and this is related to what the 95% probability actually refers to.

For brevity, let us define some quantities. Let H mean the *hypothesis* (the patient has the disease) is true, and let H' mean the hypothesis is false. Also define T to mean the disease test has been passed, and T' to mean the test comes out negative. Let $P(T|H)$ mean the probability of T occurring, given H being true. We can now define some probabilities:

- $P(H) = 0.01$. This is the probability that the hypothesis (the patient has the disease given the flu-like symptoms) is true. Therefore: $P(H') = 1 - P(H) = 0.99$.
- $P(T|H) = 0.95$. This is the probability that the test is positive, *given that the disease is present*. If 100 identical test tubes containing samples of the disease microbes are subject to the test, and 95 of them pass, then this is what, in practical terms, a test efficacy of 95% means. Note $P(T'|H) = 1 - P(T|H) = 0.05$ since one assumes the test has a binary positive/negative outcome.
- If the test is done by a reputable laboratory, they will also apply the test to 100 test tubes that are completely devoid of the disease microbes. Let's assume a 'symmetric' situation where 5 of these pass the test (i.e. the same number that fail the equivalent experiment with diseased test tubes). Hence $P(T|H') = 0.05$ and $P(T'|H') = 1 - P(T|H') = 0.95$. However, although we *might* expect some form of correlation, there is no fundamental reason why $P(T'|H') = P(T|H)$.

The notation defined above enables us to articulate the problem with the 95% statistic. What the patient really ought to want to know is the probability of the hypothesis being true, *given* a positive test, i.e. $P(H|T)$. This is *not* in general the same as $P(T|H)$. If we know $P(H|T)$ we can calculate $P(H'|T) = 1 - P(H|T)$ and hence work out the *false positive*, i.e. the problem of not having the disease given the test is passed. In this example it is $1 - 0.161 = 0.839$, i.e. the much more likely outcome. He also ought to be aware of the *false negative*, i.e. the probability of the hypothesis being true given a negative test. Surprisingly, for this example the answer is $P(H|T') = 0.001$ and hence the probability of the hypothesis being false given a negative test is $P(H'|T') = 1 - P(H|T') = 0.999$, i.e. very likely indeed. The discrepancy between $P(T|H)$ and $P(H|T)$, and the asymmetry between false positives and false negatives, results from the combination of the low probability $P(H)$ and the deviation of $P(T|H)$ and $P(T'|H')$ from unity.

Inputs to the problem are clearly $P(H) = 0.01$, $P(T|H) = 0.95$ and $P(T'|H') = 0.95$, whereas the unknown outputs are $P(H|T)$ and $P(H|T')$. We can easily relate all these quantities in a pair of tree diagrams, as shown in Fig. 3.1.

In summary, the *four* possible outcomes have probabilities:

$P(H|T) = 0.161$	$P(H|T') = 0.001$ False negative
$P(H'|T) = 0.839$ False positive	$P(H'|T') = 0.999$

$$(3.1)$$

High false positive or false negative values skew our intuitive understanding of the efficacy of tests since they modify the pair of outcomes that we normally consider, i.e. 'hypothesis true given test past' $H|T$, or 'hypothesis false given test fail' $H'|T'$. The key take-home message of this chapter is to always consider the probabilities associated with $H'|T$ and $H|T'$ as well, particularly if they are closer to unity than zero.

3.4 Tree Diagrams and Bayes' Theorem

The idea of the *tree diagram* (see Figs. 3.1 and 3.2) is to graphically represent what is actually an algebraic calculation of probability. In a tree diagram, one *multiplies* probabilities along connected branches. This is from the fundamental idea of probability being the expected fraction of a vast number of independent trials of the same system, that yield a particular outcome. Outcome H, T has probability $P(H\&T) = P(H) \times P(T|H)$ since $P(T|H)$ is the *fraction of the fraction $P(H)$* of the large population that could either be H or H'.

Although we'll bolster the idea with Venn diagrams shortly, it is perhaps logical that $P(H\&T) = P(T\&H)$. In other words, the outcome of having the disease and passing the test is the same regardless of the ordering of these phrases. However, in terms of how we construct the tree diagrams, this does make a difference. Comparing the left tree diagram in Fig. 3.1 (i.e. which corresponds to quantities we can readily calculate given our inputs) with the right tree diagram (what the patient wishes to know)

$$P(H\&T) = P(H) \times P(T|H) \tag{3.2}$$

$$P(T\&H) = P(T) \times P(H|T) \tag{3.3}$$

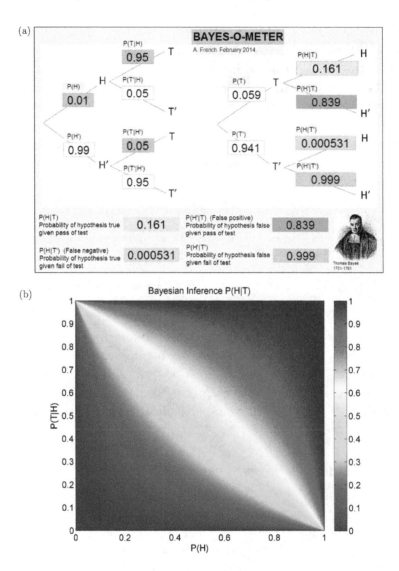

Fig. 3.1   (a) Visualisation of the tree diagrams associated with Bayesian inference in MATLAB. The orange boxes on the left are editable, whereas all others are automatically recalculated. The left diagram associates with how one might conduct a trial of a hypothesis test, given the truth or untruth of a hypothesis. The right diagram is what we actually want: the probabilities of the hypothesis being true or false given the outcome of a test. (b) $P(H|T)$ is plotted as a colour coded surface as a function of $P(T|H)$ and $P(H)$. It is assumed in this case that $P(T'|H') = P(T|H)$, hence the diagonal symmetry of the image. As $P(H)$ reduces, this means $P(T|H)$ needs to be increasing close to unity in order for $P(H|T)$ to be significant.

Hence if $P(H\&T) = P(T\&H)$

$$P(H) \times P(T|H) = P(T) \times P(H|T) \tag{3.4}$$

$$\therefore P(H|T) = \frac{P(H) \times P(T|H)}{P(T)} \tag{3.5}$$

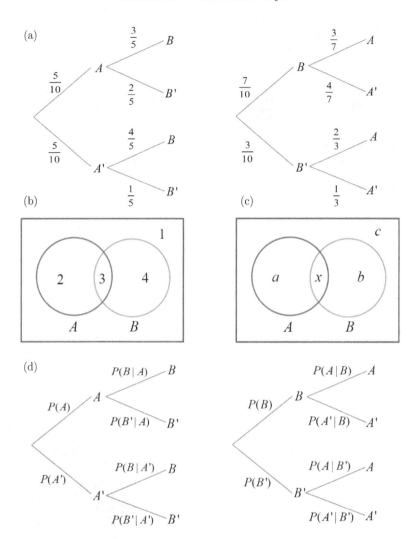

Fig. 3.2 (a) Conditional probability problem, described using a pair of tree diagrams. The full set of relationships are defined in (d). Alternatively a *Venn diagram* representation can be used, and provides an alternative mechanism to prove Bayes' theorem. An example problem involving climbers, mountaineers hillwalkers (and one couch potato) is in (b), and generalised in (c).

The statement $P(H|T) = \frac{P(H) \times P(T|H)}{P(T)}$ is what is known as *Bayes' theorem*.

Similarly $P(H'\&T') = P(T'\&H')$, hence:

$$P(H') \times P(T'|H') = P(T') \times P(H'|T') \qquad (3.6)$$

$$\therefore P(H'|T') = \frac{P(H') \times P(T'|H')}{P(T')} \qquad (3.7)$$

Now from the left 'Bayes-o-meter' diagram in Fig. 3.1(a):

$$P(T) = P(H) \times P(T|H) + P(H') \times P(T|H') \qquad (3.8)$$

$$P(T') = P(H) \times P(T'|H) + P(H') \times P(T'|H') \qquad (3.9)$$

Hence:

$$P(H|T) = \frac{P(H) \times P(T|H)}{P(H) \times P(T|H) + P(H') \times P(T|H')} \tag{3.10}$$

$$P(H'|T') = \frac{P(H') \times P(T'|H')}{P(H) \times P(T'|H) + P(H') \times P(T'|H')} \tag{3.11}$$

Our desired four outputs, $P(H|T), P(H'|T), P(H|T'), P(H'|T')$ can therefore be expressed in terms of the inputs $P(H) = 0.01$, $P(T|H) = 0.95$ and $P(T'|H') = 0.95$:

$$P(H|T) = \left(1 + \frac{P(H') \times P(T|H')}{P(H) \times P(T|H)}\right)^{-1} \tag{3.12}$$

$$P(H'|T') = \left(1 + \frac{P(H) \times P(T'|H)}{P(H') \times P(T'|H')}\right)^{-1} \tag{3.13}$$

$$P(H'|T) = 1 - P(H|T) \tag{3.14}$$

$$P(H|T') = 1 - P(H'|T') \tag{3.15}$$

Let's run the numbers using this example to make sure it agrees with Fig. 3.1(a). The 'Bayes-o-meter' is a visualisation of the tree-diagram calculations above using MATLAB. The inputs $P(H) = 0.01$, $P(T|H) = 0.95$ and $P(T'|H') = 0.95$ correspond to the orange boxes in the left tree diagram. These can be edited, and the outputs (the other four colours) are automatically recalculated.

$$P(H|T) = \left(1 + \frac{0.99 \times 0.05}{0.01 \times 0.95}\right)^{-1} = 0.161 \tag{3.16}$$

$$P(H'|T') = \left(1 + \frac{0.01 \times 0.05}{0.99 \times 0.95}\right)^{-1} = 0.999 \tag{3.17}$$

$$P(H'|T) = 1 - P(H|T) = 0.839 \tag{3.18}$$

$$P(H|T') = 1 - P(H'|T) = 0.001 \tag{3.19}$$

3.5 A Proof of Bayes' Theorem Using Venn Diagrams

For certain situations, a *Venn diagram* might prove a more intuitive point of departure (than say a tree diagram) for probability calculations involving random selections from *sets* of objects which have shared characteristics. In the example given in Fig. 3.2 there are two populations, A and B, which describe the attendees of a meeting of rock climbers (A) and mountain walkers (B). The intersection[1] of the

---

[1] An *intersection* of two sets A, B is $A \cap B$. Alternatively, the *union* of the two sets is: $A \cup B$. The number of elements in any set, e.g. A, is given by $n(A)$.

two sets are the people who combine their climbing and mountain walking skills, and therefore belong to both populations. We shall call these 'mountaineers'. In the example we have five climbers, i.e. $n(A) = 5$, 7 mountain walkers, i.e. $n(B) = 7$, and an overlap of $n(A \cap B) = 3$ who are members of sets A and B. Also at the meeting, is a friend of one of the climbers who really can't stand either climbing or mountain walking, but hears there may be free beer i.e. $n(A' \cap B') = 1$. The dash notation means 'not' or 'does not belong to'. In total, 10 people are at the meeting. It turns out the free beer is only to be available to one lucky person, and this is to be chosen at random via a raffle. Everybody has the same chance of winning. The probabilities of obtaining the beer for each grouping of people are easily determined from inspection of the Venn diagram in Fig. 3.2(b):

$$P(A) = \tfrac{5}{10} = \tfrac{1}{2}, \quad P(B) = \tfrac{7}{10} \tag{3.20}$$

$$P(B|A) = \tfrac{3}{5}, \quad P(A|B) = \tfrac{3}{7} \tag{3.21}$$

$$P(A \ \& \ B) = P(A \cap B) = \tfrac{3}{10} \tag{3.22}$$

$$P(A \text{ or } B) = P(A \cup B) = \tfrac{9}{10} \tag{3.23}$$

It is easy to show that Bayes' theorem holds, which in this case would be: $P(A) \times P(B|A) = P(B) \times P(A|B)$

$$P(A) \times P(B|A) = \tfrac{1}{2} \times \tfrac{3}{5} = \tfrac{3}{10} \tag{3.24}$$

$$P(B) \times P(A|B) = \tfrac{7}{10} \times \tfrac{3}{7} = \tfrac{3}{10} \tag{3.25}$$

One can readily generalize, using $n(A) = a + x$, $n(B) = b + x$, $n(A \cap B) = x$, $n(A' \cap B') = c$, i.e. with 'everything' ε corresponding to the number of elements $n(\varepsilon) = a + b + x + c$ (see Fig. 3.2). The probabilities are:

$$P(A) = \frac{a + x}{a + b + x + c} \tag{3.26}$$

$$P(B) = \frac{b + x}{a + b + x + c} \tag{3.27}$$

$$P(B|A) = \frac{x}{a + x} \tag{3.28}$$

$$P(A|B) = \frac{x}{b + x} \tag{3.29}$$

$$P(A \ \& \ B) = P(A \cap B) = \frac{x}{a + b + x + c} \tag{3.30}$$

$$P(A \text{ or } B) = P(A \cup B) = \frac{a + b + x}{a + b + x + c} \tag{3.31}$$

Hence:

$$P(A) \times P(B|A) = \frac{a+x}{a+b+x+c} \times \frac{x}{a+x} = \frac{x}{a+b+x+c} \qquad (3.32)$$

$$P(B) \times P(A|B) = \frac{b+x}{a+b+x+c} \times \frac{x}{b+x} = \frac{x}{a+b+x+c} \qquad (3.33)$$

i.e. proving Bayes' theorem $P(A) \times P(B|A) = P(B) \times P(A|B)$ is generally true, but without using tree-diagram ideas.

3.6 An Inversion of Bayes' Formula

To return to the disease-testing example, one might wish to know how accurate the test should be, i.e. the value of $y = P(T|H)$, in order to yield $p = P(H|T)$ to be a suitably high probability, e.g. $p = 0.9$. In summary, define:

$$x = P(H) \qquad (3.34)$$

$$y = P(T|H) \qquad (3.35)$$

$$k = P(T'|H') \qquad (3.36)$$

Using the formulae above:

$$P(H|T) = \left(1 + \frac{P(H') \times P(T|H')}{P(H) \times P(T|H)}\right)^{-1}$$

Hence:

$$p = \left(1 + \frac{(1-x)(1-k)}{xy}\right)^{-1}$$

$$\frac{1}{p} - 1 = \frac{(1-x)(1-k)}{xy}$$

$$\therefore y = \frac{(1-x)(1-k)\,p}{x\,(1-p)} \qquad (3.37)$$

Since the parameters correspond to real probabilities which are in the range [0,1]:

$$0 \le k \le 1 \qquad (3.38)$$

$$0 \le y \le 1 \qquad (3.39)$$

Figure 3.3 yields some interesting conclusions. As $P(H)$ becomes closer to unity, the requirement for a super-accurate disease test diminishes, indeed to zero when $P(H) = 1$. A problem with this is one might expect some form of correlation between $k = P(T'|H')$ and $y = P(T|H)$ rather than k being fixed at 0.95. For an alternative view, let us consider the extreme correlation, i.e. what we defined as a 'symmetric' situation, $k = y$, i.e. the probability of a negative result given no disease is the

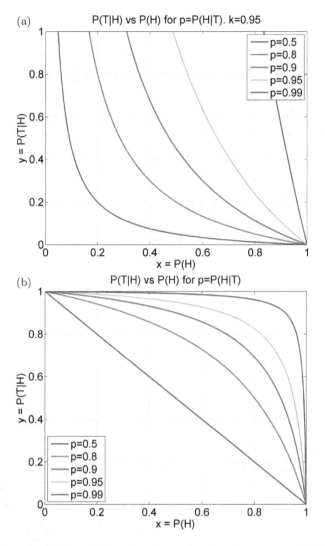

Fig. 3.3   (a) An inversion of Bayes' formula plots a family of curves $y = P(T|H) = \frac{(1-x)(1-k)p}{x(1-p)}$ where $p = P(H|T)$ and $x = P(H)$ and $k = P(T'|H')$. In this example $k = 0.95$. This demonstrates the sensitive dependence of 'test efficacy' $P(T|H)$ upon $P(H)$, particularly as $p \to 1$. (b) An alternative set of curves if instead $y = k$, i.e. $P(T|H) = P(T'|H')$.

same as the probability of a positive result given disease is definitely present. In this situation we need a slightly different rearrangement to find y:

$$\frac{1}{p} - 1 = \frac{(1-x)(1-y)}{xy} \tag{3.40}$$

$$\frac{1-p}{p} = \frac{1-x}{x}\left(\frac{1}{y} - 1\right) \tag{3.41}$$

$$\frac{1-p}{p}\frac{x}{1-x} + 1 = \frac{1}{y} \tag{3.42}$$

$$\therefore y = \left(\frac{1-p}{p} \frac{x}{1-x} + 1 \right)^{-1} \tag{3.43}$$

The relationships in Fig. 3.1 are possibly more intuitive than in Fig. 3.3, that is, one expects $y = P(T|H)$ to remain very close to unity to achieve $p = 0.9$ (or higher), until $P(H)$ itself becomes close to unity. An interesting special case is when $p = 0.5$, which implies the taking of the test for the disease is no better than the flipping of a fair coin. In this situation:

$$y = \left(\frac{1-0.5}{0.5} \frac{x}{1-x} + 1 \right)^{-1} \tag{3.44}$$

$$y = \left(\frac{x}{1-x} + 1 \right)^{-1} \tag{3.45}$$

$$y = \left(\frac{x+1-x}{1-x} \right)^{-1} \tag{3.46}$$

$$y = 1 - x \tag{3.47}$$

i.e. we expect a linear relationship between x and y, which means the probability of passing the test is $P(T|H) = 1 - P(H)$.

3.7 A Simulation Involving Bayesian Inference

3.7.1 *Bayes urns its keep, making use of the binomial distribution*

Of the most exciting aspects of 'Bayesian inference' is how one can recalculate the probability of a particular hypothesis $P(H|T)$ in the light of new evidence. The mathematics we have developed above is broadly applicable, and we can interpret T as being some form of generic test of hypothesis H. Repeated testing is effectively an *information* gathering exercise, and this means we can use a calculation system based upon Bayesian ideas to make *decisions* that a hypothesis is 'likely to be true', once the probability $P(H|T)$ exceeds some acceptable threshold (e.g. 95%). Although the author would not recommend trying these methods in a casino (the owners and the other players will take a very dim view), the ideas of Bayesian statistics can be used to work of the odds of the hands of the other players in card games, given the cumulative information revealed as the game progresses.[2]

To further illustrate the power of Bayesian inference, we shall investigate a much simpler system, but one that offers more knowledge as additional tests are conducted.

---

[2]The idea of 'card counting', most infamously described in Ben Mezrich's book *Bringing Down the House*, almost certainly involves a more human-centric method of computation than the computer-suited ideas presented here. The book describes the story of six MIT students who develop a collaborative system to win more than three million dollars at the Blackjack tables of Las Vegas casinos.

The scenario is as follows: There are two large opaque ceramic urns containing thousands of small marbles. These are either red R or white W in colour. Urn A has proportion p of red marbles whereas Urn B has proportion q of red marbles. The idea is one urn is chosen at random, and N marbles are taken out and their colour noted. Strictly speaking, the marbles should be replaced and shaken around in the urn between draws, but the number of marbles in the urn could be deemed large enough that the change of p or q, following a draw of N marbles, can be neglected. The question is, what is the probability that the Urn is A, given N draws? How might this vary depending on the relationship between p and q? (see Fig. 3.4). This problem can be implemented via a computer simulation which produces ball colours appropriate to the distribution of the Urn. As balls are chosen, the odds of the Urn being A or B given the history of the measurements (i.e. how many x of N balls are red) can be calculated (see Fig. 3.5).

Since the red, white marble distribution of both urns doesn't change, we can make a fair assumption that each draw is completely independent of the other. N marbles could be drawn one after the other, or perhaps as a handful if N is small compared to the number of marbles in each urn. Let's assume we draw x red marbles and $N - x$ white marbles. If the urn is A, the probability of this selection of red and white is the probability of a sequence of x reds, then $N - x$ whites multiplied by the number of ways of arranging this sequence. The result is known as the *Binomial Distribution*:

$$P(x|A) = \frac{N!}{(N-x)!x!}p^x(1-p)^{N-x} \tag{3.48}$$

The equivalent probability for Urn B is:

$$P(x|B) = \frac{N!}{(N-x)!x!}q^x(1-q)^{N-x} \tag{3.49}$$

A (normalised) histogram of 1,000,000 samples from a Binomial distribution with $p = 0.3$ and $N = 10$ is plotted in Fig. 3.4(e). The red stars are the Binomial distribution probabilities $P(x) = \frac{10!}{(10-x)!x!}(0.3)^x(0.7)^{10-x}$. Normalisation means a scaling of the frequency of samples such that the total is 1. In the limit of infinite samples, the normalised histogram should tend towards the probability distribution.

The problem in our marble-picking scenario is we don't know which urn we have chosen. What we want to calculate are the probabilities of $P(A|x)$ and hence $P(B|x) = 1 - P(A|x)$. However, the Binomial distribution only allows us to find $P(x|A)$ and $P(x|B)$. We can achieve what we actually want by using *Bayes' theorem*:

$$P(x)P(A|x) = P(A)P(x|A) \tag{3.50}$$

Now, with reference to the tree-diagram in Fig. 3.1:

$$P(x) = P(A)P(x|A) + P(B)P(x|B) \tag{3.51}$$

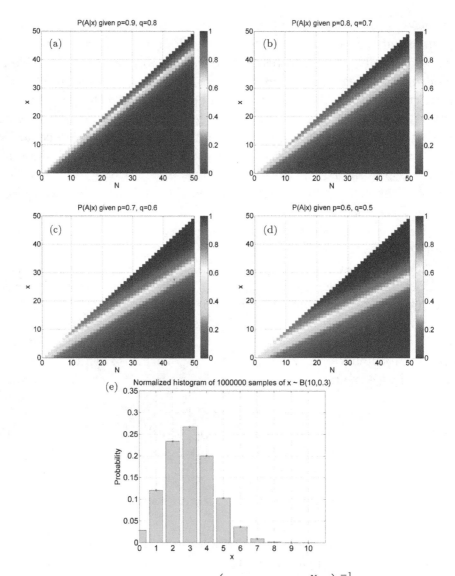

Fig. 3.4   A coloured surface plot of $P(A|x) = \left(1 + \left(\frac{q}{p}\right)^x \left(\frac{1-q}{1-p}\right)^{N-x}\right)^{-1}$ is illustrated for x and N values over the range 0 to 50. x is the cumulative number of red marbles picked, out of N picks. The probability $P(A|x)$ corresponds to the colour scale, with 0 dark blue and 1 dark red. (a)–(d) are the plots for different pairings of p, q, values. These correspond to the proportion of red marbles in urns A and B, respectively. (e) is a plot of the normalised histogram of one million samples from a Binomial distribution, with $p = 0.3$ and $N = 10$. The red stars are the Binomial distribution probabilities.

and if the urn is chosen at random:

$$P(A) = P(B) = 0.5 \tag{3.52}$$

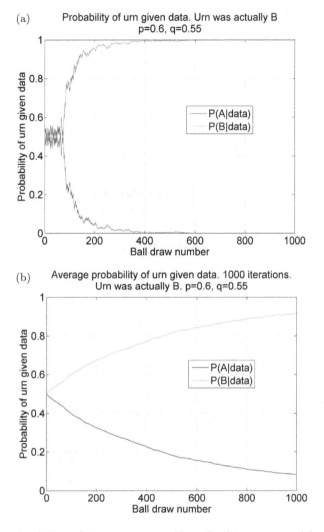

Fig. 3.5 Computer simulation of the two-urns problem. In this case, urn A has 60% red marbles and urn B has 55% red marbles, and picks from each urn are chosen at random on this basis. (a) represents the probabilities of urn A (blue) and urn B (green) given N draws from a, randomly selected, urn. In this case the urn was actually B. The jagged progression of probabilities is to be expected given the random nature of the picks. (b) is a mean average of these probability traces over 1,000 iterations. The more gentle progression to certainty implies the (steeper) trace in (a) is perhaps rather a 'lucky' run.

Hence:

$$P(A|x) = \frac{P(A)P(x|A)}{P(A)P(x|A) + P(B)P(x|B)} \qquad (3.53)$$

$$P(A|x) = \left(1 + \frac{P(B)P(x|B)}{P(A)P(x|A)}\right)^{-1} \qquad (3.54)$$

$$P(A|x) = \left(1 + \frac{P(x|B)}{P(x|A)}\right)^{-1} \tag{3.55}$$

$$P(A|x) = \left(1 + \frac{q^x(1-q)^{N-x}}{p^x(1-p)^{N-x}}\right)^{-1} \tag{3.56}$$

$$\therefore P(A|x) = \left(1 + \left(\frac{q}{p}\right)^x \left(\frac{1-q}{1-p}\right)^{N-x}\right)^{-1} \tag{3.57}$$

Figure 3.4 evaluates the formula for $P(A|x)$ above. The dependence upon both x and N naturally leads to a coloured surface representation, for a given pair of p, q values. The sequence of surfaces for $(p, q) = (0.9, 0.8), (0.8, 0.7), (0.7, 0.6), (0.6, 0.5)$ in Fig. 3.4 demonstrate the trend that the steepness of the transition from low to high probabilities decreases as the ratio $\frac{p}{q}$ increases. When both p and q are relatively high, there is a fairly sharp increase from effectively zero to near unity. This might intuitively seem like a strange result, but if both p and q are close to unity (and $p > q$), it is very unlikely x will be significantly less than N, regardless of the urn chosen. If this *does* happen though, the Bayesian result will predict Urn B with high certainty.

3.7.2 *A computer simulation of the binary urn problem*

Figure 3.5 approaches the urn-choosing problem from a different perspective. Rather than mapping out the possible probabilities as in Fig. 3.4, this simulation aims to model the scenario in a more direct way. An urn (either A or B) is chosen at random, with equal probability, and up to 1,000 marbles are drawn one by one, and their colour (red or white) is noted. Given the large number of marbles, it is assumed the marbles are replaced after each draw and some form of shuffling occurs continuously. In the simulation in Fig. 3.5, urn B was actually chosen, and $p = 0.6$ and $q = 0.55$, i.e. each urn has a broadly similar proportion of red marbles.

In the MATLAB simulation described in Fig. 3.5, the sequence of red or white marbles is defined by firstly creating a vector of 1,000 random numbers, uniformly distributed between 0 and 1. The built-in `rand(1,1000)` command will achieve this in MATLAB. Those numbers $\leq p$ (if Urn A is chosen) or $\leq q$ (if urn B is chosen) are set to be ones (i.e. the red ball picks), and zeros otherwise. The cumulative sum therefore yields x, out of N picks. In this case N is now a vector with elements $1, 2, 3 \ldots, 1000$. If the (binary) pick sequence is $[0, 0, 1, 1, 1, 0, 1, 1 \ldots]$ then x would be $x = [0, 0, 1, 2, 3, 3, 4, 5, \ldots]$ and $N = [1, 2, 3, 4, 5, 6, 7, 8, \ldots]$.

The formulae:

$$P(A|x) = \left(1 + \left(\frac{q}{p}\right)^x \left(\frac{1-q}{1-p}\right)^{N-x}\right)^{-1}$$

and

$$P(B|x) = 1 - P(A|x) \tag{3.58}$$

are then used to calculate the probability of urns A and B (respectively) given a sequence of x red marbles recorded out of N draws. Figure 3.5(a) illustrates one iteration of this model run, and the jagged lines illustrate the random nature of the process. However, after about 200 draws there is about a 95% chance that the urn is B, and 5% that it is A, given the (binary) sequence. After 400 draws the correct choice is close to unity, and beyond 600 draws $P(B|x) = 1.00$, to the visual precision of the plot. Is this model run representative? Or is it a statistical abberation? To help answer this question, 1,000 iterations of the model were run in a loop, and a mean average of the $P(A|x)$ probabilities calculated for each draw over all 1,000 iterations. A plot of these averages is illustrated in Fig. 3.5(b), and suggests a rather less steep progression, on average, to certainty that urn B was chosen.

A decision-making machine based upon these Bayesian ideas might be designed on the basis of this more conservative estimate of how many picks are required, in order to achieve an acceptable level of probability. For a team of 'legally flexible' students taking on a casino, it might be rather important that this number is not too high, as the risk of detection will certainly rise with the length of time the student is engaged with a particular game of chance. In our binary urn scenario, the efficiency of the Bayesian method is, quite literally, subject to the luck of the draw.

3.8 The Influence of Inference

"When you have eliminated the impossible, whatever remains, however improbable, must be the truth." — Arthur Conan Doyle, *The Sign of Four*

"It is a capital mistake to theorize before one has data. Insensibly one begins to twist facts to suit theories, instead of theories to suit facts." — Arthur Conan Doyle, *A Scandal in Bohemia*

This chapter builds upon the ideas of the *Stochastic Eyam* model in the previous chapter, and demonstrates how we can incorporate random chance and probability in computer simulations. A close study of systems involving conditional probability and Bayesian inference remind us of the fundamental logical machinery of the *Scientific Method*. What is the chance a theory is true, given a set of observations? The basic methodology of Science is to follow the following cyclical process:

(1) Make an observation, ideally quantitative. For example, sunrise on Wednesday 6 January 2020 in Winchester, UK is 08:08 (GMT).
(2) Propose a theory, as simple and generally applicable as possible, that describes correlations between observations, typically in the form of equations. For the example above, a celestial model of the motion of the Earth relative to the Sun,

incorporating the tilt and rotation of the Earth about its axis, the latitude and longitude of Winchester, etc.

(3) Test theoretical predictions by doing an unbiased experiment. In our sunrise example, getting up early with an accurate clock and recording the time of sunrise, and then for many subsequent mornings, having run the model for later dates beforehand and writing down the predictions. To avoid systematic error, the sun must not be impeded by topography, nearby buildings or particularly thick cloud cover.

(4) If there is deviation from theoretical predictions, and experimental errors cannot be identified, it is the theory which must be modified. Elegant but flawed theories must be rejected in the light of a mismatch with reality, regardless of the reputation of the proponent of the theory. Make more observations, refine the theory, and continue the cycle.

In most modern situations, complete rejection of a theory following the discovery of anomalies is rarely necessary. Augmentation is more typical. A classic example is the need to modify Newton's inverse-square law of the gravitational force between two masses in the light of close observations of the orbit of Mercury, and the bending of light by massive stars. These observations, coupled with the necessary modifications to the equations of dynamics to take into account the constancy of the speed of light for all observers in relative motion, led Albert Einstein and his contemporaries to propose a radical new theory of gravity, *General Relativity*. However, in the low speed, low force limit of our common experience of the solar system, the equations of General Relativity produce the same orbit descriptions as Newton's laws. This is a wonderful feature of the Scientific Method. Once a theory has been found that reliably explains many observations, it must still be valid (in a certain restricted sense) even when a more advanced theory has been discovered. Although we might talk of revolutions, good Science builds upon the most solid of foundations, and new Science adds rather than replaces.

Chapter 4

May's Chaotic Bunnies

4.1 What Is Chaos?

Dynamics, the physics of motion, provides us with equations which can be used to predict the future position of objects if we know (i) their present *position* and *momentum* and (ii) the *forces* which act on each object. This is effectively a statement of Newton's second law: *the vector sum of force = the rate of change of momentum*. The French polymath Pierre Simon Laplace (1749–1827) applied this idea to the largest of scales: if somehow all the information about the current dynamical state of the Universe could be known, it would be possible to know with complete precision both the past *and* the future. Unfortunately it appears this grand philosophical extrapolation of Laplace is flawed in three fundamental ways. Firstly, our measurement uncertainties in position Δx and momentum Δp must obey the *Uncertainty Principle* of Quantum Mechanics, that is $\Delta p \Delta x \geqslant \frac{1}{2}\hbar$ where 'h-bar', Planck's constant divided by 2π, is $\hbar = \frac{1}{2\pi} \times 6.626 \times 10^{-34}$ Js. Energy ΔE and time Δt uncertainties obey the same inequality. So we cannot know, simultaneously, the position and momentum of *any* particle with total precision, let alone a Universe full of interacting particles.

This fundamental Quantum mechanics measurement problem is compounded by our ability to model *analytically* the interplay between forces and resulting position and momentum of a given object. The motion of a Newtonian apple or the orbit of Earth about the Sun are very special cases of dynamical systems that we can *solve*. In other words, we can determine a mathematical expression for position and velocity vs time given a precise expression of Newton's second law. Mathematically, this amounts to a single integral to yield velocity vs time from acceleration, and then a further integral to give position vs time from velocity. Very easy to say, but not always so easy to do in practice. Although it may seem like the multi-inch textbooks of mathematics and physics must contain the answers to all possible problems, the reality is that only a select few systems can be solved analytically. Perhaps the most stark example is that of the gravitational interaction between masses (and since the force law is mathematically the same, the electromagnetic interaction between point charges[1]). It is possible to solve a two-body system precisely, but a three-body system? No. Anything beyond the inverse-square gravitational interaction between two masses requires an approximate numerical method to solve. Much like the *Euler* scheme used to solve the *Eyam equations* in Chapter 2, these rely on small, discrete, but *finite* time steps. Within this interval, the force laws can be used to predict what will happen next. An approximation is employed, e.g. constant acceleration during the discrete time interval Δt (this is called the *Verlet* method), or more precise but computationally more intensive schemes such as *Runge–Kutta* methods. The idea is that the numerical recipes become more accurate as the time interval is reduced. However, reducing Δt increases the number of (iterative) calculations required, and a time-step of zero would mean an *infinite* number of iterations. A fully precise prediction of past or future is therefore impossible to calculate, even if you could know the status quo exactly.

The third problem relates to how errors in position and momentum grow with time. The celestial clockwork of the orbits of planets in the solar system is very different from the variation of wind velocity, air pressure, temperature and humidity in the troposphere of the Earth. The interaction between these variables is such that future behaviour can be *highly sensitive to initial conditions*. Since we can only know the initial system state approximately, errors can amplify such that even the super-computers of the UK Meteorological Office can predict the weather with certainty only a few days ahead. The position of pool balls after a break is another good example of sensitivity to initial conditions, and the difficulty of making predictions via an approximate numerical solver. A tiny change in how the white ball is struck will result in a wildly different final position of the balls. Indeed the gameplay of pool *relies* on this property, as the interaction between pool balls during a break effectively randomises the positions of balls at the start of the game. Pool would be somewhat boring (and unfair) without it.

Robert Shaw of the "Santa Cruz Chaos Cabal" in 1970–1980s California gives a nice definition of *Chaos*. It is not randomness *per se*, but seemingly random

---

[1]Although in the electrical case, there is the additional phenomenon of *magnetism* which results from moving charges (i.e. currents), and indeed the production of electromagnetic *radiation* which results when charges accelerate. Does gravity work in the same way too? Modern orthodoxy is the idea that the effect of gravity manifests from a *curvature* of space (and time) in the vicinity of mass and energy. This is Einstein's theory of General Relativity. Although I think 'gravitomagnetism' is regarded as rather a fringe theory, recent experiments such as LIGO have shown that gravitational radiation does indeed exist. Detection requires truly exquisite precision, but it is possible to measure the ripples in spacetime following the merger of extreme, and extra-galactic, objects such as pairs of inspiralling black holes [1].

behaviour resulting from the known (i.e. *deterministic*) interactions between objects. To quote from Shaw (1986) [43] and Gleick (1998) [17]:

> "Simple deterministic systems with only a few elements can generate random behaviour. The randomness is fundamental; gathering more information does not make it go away. Randomness generated in this way has come to be called chaos."

4.2 The Link Between Chaos and Nonlinearity

Edward Lorenz (1917–2008), whose *strange-attractor* we will meet in Chapter 5, devised a system of equations for modelling weather.

> "... it is found that non-periodic solutions are ordinarily unstable with respect to small modifications, so that slightly differing initial states can evolve into considerably different states."

This means if I change pressure (or any other meteorological variable) by even a tiny amount in a weather model, the effect may be profound after a relatively short time. Lorenz coined this the *Butterfly Effect*, i.e. "does the flap of a butterfly's wings in Brazil set off a tornado in Texas?" Lorenz's famous phrase is doubly apt, since the shape of his attractor might also resemble a (albeit rather abstract) butterfly! The Butterfly Effect may also have rather deep philosophical implications regarding *causality*, as it might be computationally impossible to determine whether an intervention in a moderately complicated system is the *primary cause* of some future outcome. Perhaps an appraisal of corporate and political management should be more mindful of this idea, and measures of success should be determined in terms of metrics of good stewardship, rather than a dubious linkage of personal intervention to socio-economic outcome.

Systems that are particularly sensitive to initial conditions, and also are less likely to have analytical solutions, are those which typically contain *nonlinear terms*. Linearity means a 'straight-line' or 'constant gradient' relationship. The force to extend a spring by a relatively small fraction of its original length (i.e. Hookes' law) and the relationship between potential difference and current in a fixed resistor (i.e. Ohm's law) are both linear. The idea can be extended to differential equations. In radioactivity, the activity of a sample (i.e. the number of decays per second dN/dt) is proportional to the number of radioactive atoms N that have not yet decayed. The equation $\frac{dN}{dt} = -\lambda N$ is linear, and solvable. (The solution is $N = N_0 e^{-\lambda t}$.) An equation like $\frac{dy}{dx} = xy$ may not be linear, but it solvable since it is *separable* in terms of the variables x and y, i.e. $\int \frac{1}{y} dy = \int x dx \Rightarrow y = y_0 e^{\frac{1}{2}x^2}$. However a differential equation such $\frac{dy}{dx} = \sin(xy)$ is not linear, and not separable either. It requires a numerical method to solve.

Much of the canon of mathematics (certainly pre-University mathematics) is related to the study of linear equations, and more broadly, equations that we can solve analytically. Systems which require numerical methods are typically shunned and indeed numerical methods are barely present in most (UK) schemes of work such as A-Levels. However, this is also a problem in the real world beyond school. Ignoring nonlinearity and numerical methods is counter to the reality that mathematics is applied to, and may give students a false sense of what mathematics can achieve. One might take the economic chaos of recent decades being a salutary example. To paraphrase Stanislaw Ulam (1909–1984), quoted in James Gleick's *Chaos* [17], much of mathematics (and theoretical physics too) is like zoology being the principle study of elephants, with a short back-of-the-textbook appendix on non-elephants. The study of nonlinearity is, quite literally, the non-elephant in the room (!) May devised a wonderfully simple model, that we will shortly discuss, as an impassioned plea for nonlinear equations and numerical methods to be included in pre-University courses. Not only does it reveal the limitations of our analytical mathematical methods, but also reveals exquisite beauty and complexity inherent in even the simplest of mathematical expressions. We will elaborate on this idea considerably in Chapter 8 entitled *Exploring Julia's Fractals*.

4.3 May's Bifurcating Bunnies

Robert May (the Australian scientist, Professor of Sydney, Princeton, Imperial College and Oxford universities, crossbench member of the UK House of Lords and 59th President of the Royal Society, but *not* the creator of Rudolph the Red Nosed Reindeer) proposed a beautifully simple, but nonlinear, model of population growth that can be used to illustrate chaotic behaviour, and indeed the progression towards it. Unlike Lorenz's system, May provides a single *logistic* equation, that relates the fraction x_{n+1} of a total permissible population that will be alive next year, given the population this year x_n, and a *growth parameter* r. The fixed timestep of 1 year is built in, no approximation of a differential equation is needed. May's equation is therefore a very sensible first recipe in a computational cookbook of chaos.

4.3.1 *May's logistic equation*

Assume an ecosystem can support a maximum number of rabbits. Let x be the fraction of this maximum at year n. To account for reproduction, next year's population is proportional to the previous, i.e. $x_{n+1} \propto x_n$. To account for the population-limiting effect due to a finite resource of food, shelter etc, let next year's population be also proportional to the fraction of the maximum population as yet unfilled, i.e. $x_{n+1} \propto (1 - x_n)$. Hence next year's population is predicted by the following logistic equation, which combines these two factors:

$$x_{n+1} = rx_n(1 - x_n) \tag{4.1}$$

To investigate the variation of x with numbers of years n we shall start with a spreadsheet, as illustrated in Fig. 4.1. Eighteen iterations are represented by

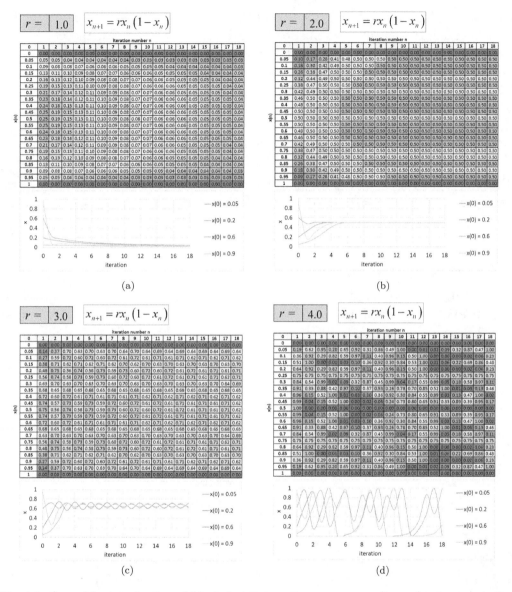

Fig. 4.1 Spreadsheet evaluation of May's *logistic equation* $x_{n+1} = rx_n(1 - x_n)$, where x_n is the fraction of a maximum permissible population at year n. This means $0 \leq x \leq 1$. (a)–(d) apply the same equation, but with different growth parameters r, over 18 iterations (the columns) based upon starting values (x_0) varying from 0 to 1 in 0.05 steps (the rows). Extinction occurs if $r \leq 1$, a single asymptotic value is attained if $1 \leq r < 3$, and bifurcations (i.e. alternate cycles of two possible asymptotic populations) start to occur when $r > 3$. When $r = 4$, the variation of x with n is fully chaotic.

consecutive columns, and the rows correspond to starting conditions varying from $x_0 = 0$ to $x_0 = 1$ in steps of 0.05. The growth parameter r is fixed at 1, 2, 3, 4 and corresponds to the subplots (a), (b), (c), (d). The difference in characteristic 'trajectories', given different growth parameters, is significant. For $r = 1$, there is some form of rabbit apocalypse, and all populations become extinct, regardless of the starting conditions. When $r = 2$, and with the exception of $x_0 = 0$ or 1, which clearly result in immediate extinction given $x_{n+1} = rx_n(1 - x_n)$, the population tends to a fixed value, regardless of the initial conditions. When $r = 3$, this asymptotic behaviour changes to an alternate-year oscillation between a pair of possible high, low values of proportion x of the maximum rabbit population. However, when $r = 4$, this regularity breaks down to a seemingly random connection between the population at year n and x_0. In other words, increasing the growth parameter leads us from the regularity of a fixed steady-state solution, to oscillations, to chaos.

An implementation of the same scenario using MATLAB is illustrated in Fig. 4.2(a). In this case, a more fine-grained 'spreadsheet' is worked *in-situ*, with 100 randomly chosen starting values and 50 iterations. Coloured dots represent the 100 different 'trajectories'.

4.3.2 *A logistic map of May bifurcations*

In addition to demonstrating a transition from simplicity to complexity, an analysis of May's equation introduces many other geometrical (i.e. 'picture based') tools that are characteristic of what is often packaged as *Chaos Theory*. If $x(t)$ is chaotic, it doesn't really make much sense attempting to analyse a given time series. Sensitivity to initial conditions means that a small variation in $x(0)$ will result in a completely different trajectory. It is better to overlay *all* the possible trajectories, and see if patterns emerge.

A *logistic map* for $x_{n+1} = rx_n(1 - x_n)$ can be constructed using a computer program with the following instructions:

- For every growth parameter r_n (in N steps between limits of $r_{\min}$ and $r_{\max}$), determine N possible starting values[2] of x_0, randomly chosen from range $0 \leq x_0 \leq 1$.
- Apply $x_{m+1} = r_n x_m(1 - x_m)$ for each growth parameter r_n, for M iterations, and for each of the N distinct starting positions. Keep going for another M iterations, and plot these (r, x) points on a two-dimensional Cartesian graph with r being the horizontal axis and x being the vertical axis.

  In a language like MATLAB, this means creating $N \times M$ arrays for r and x. The array for r will be the same column vector repeated over M columns. MATLAB can then be used to efficiently render a scatter plot of r against x. This scatterplot is illustrated in Fig. 4.2(b). For the plot, $N = 2000, M = 2000$. This gives a much clearer picture of the scenarios hinted at above.
- For $r \leq 1$, the asymptotic population tends to zero.

---

[2]There is no fundamental reason why the number N of growth parameter r values should equal the number of different starting conditions for x, but to keep things simple I shall set this to be so.

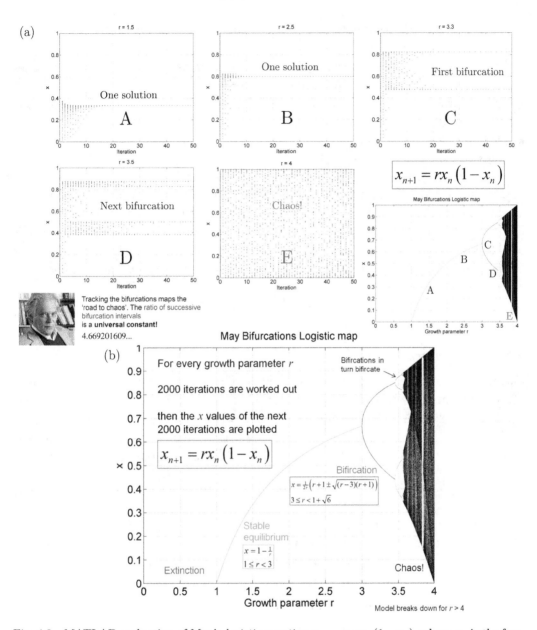

Fig. 4.2   MATLAB evaluation of May's *logistic equation* $x_{n+1} = rx_n(1-x_n)$, where x_n is the fraction of a maximum permissible population at year n. This means $0 \leq x \leq 1$. Figures A to E within subplot (a) apply the same equation, but with different growth parameters r, over 50 iterations based upon 100 starting values (x_0) randomly chosen from the interval [0,1]. Extinction occurs if $r \leq 1$, a single asymptotic value is attained if $1 \leq r < 3$, and bifurcations (i.e. alternate cycles of two possible asymptotic populations) start to occur when $r > 3$. When $r = 4$, the variation of x with n is fully chaotic. A map of all possible x values, plotted against r, is illustrated in subplot (b). 2,000 randomly chosen initial x values from the interval [0,1] are subjected to May's equation, and the x values for iterations 2,001 to 4,000 are all plotted. This is then repeated for different r values. For the range $1 + \sqrt{6} \leq r < 4$, a progressive cascade of bifurcations, with successive bifurcation intervals tending towards the *Feigenbaum Number* ratio: $\lim_{n \to \infty} \left(\frac{r_n - r_{n-1}}{r_{n+1} - r_n} \right) = 4.669201609\ldots$

- For $1 \leq r < 3$, the population tends to a single value, which is 0.5 when $r = 2$ and close to $\frac{2}{3}$ when $r = 3$.
- After $r > 3$ there is a bifurcation, i.e. the population oscillates between two values $x_{\pm}$. When $r \approx 3.45$, each bifurcation *itself* bifurcates.
- As r approaches 4, there has been an infinite cascade of bifurcations, and the trajectory of x has become chaotic.
- The model diverges when $r > 4$, i.e. we get x values outside range [0,1] and therefore the model is not applicable.

Mitchell Feigenbaum (1944–2019) discovered the bifurcation branching effect was a *universal phenomenon*, i.e. not just related to May's logistic equation (see Strogatz [49, pp. 353–387]). The ratio of successively smaller 'bifurcation intervals', in May's case the difference in r values between subsequent bifurcation points, always tends to the *same constant*, an irrational number known as the *Feigenbaum Number*.

$$\lim_{n \to \infty} \left(\frac{r_n - r_{n-1}}{r_{n+1} - r_n} \right) = 4.669201609\ldots \tag{4.2}$$

4.3.3 *Why must $r \leq 4$?*

Our model requires x to be in the range [0,1]. This means $rx(1-x) \geq 0$, and since $r > 0$, this implies $0 \leq x \leq 1$, which is just what we require. If this is at all unclear, think about sketching the inverted parabola $y = x(1-x)$. It is above the x axis in the range $0 \leq x \leq 1$. Now consider $rx(1-x) \leq 1$. We can write this as a quadratic inequality $x - x^2 - \frac{1}{r} \leq 0$, or $0 \leq x^2 - x + \frac{1}{r}$, which by completing the square is:

$$\left(x - \tfrac{1}{2} \right)^2 - \tfrac{1}{4} + \tfrac{1}{r} \geq 0 \tag{4.3}$$

For this to be always true for any value of x, $-\frac{1}{4} + \frac{1}{r} \geq 0$ since $\left(x - \frac{1}{2} \right)^2$ is always positive and has a minimum value of zero at $x = 1/2$, i.e. the vertex of the quadratic $y = \left(x - \frac{1}{2} \right)^2 - \frac{1}{4} + \frac{1}{r}$. Hence $\frac{1}{r} \geq \frac{1}{4} \Rightarrow r \leq 4$ as required.

4.3.4 *Solving for a single asymptotic value of x*

Consider the possibility of a steady-state, i.e. $x_{n+1} = x_n$. We can solve for this by stating $x = rx(1-x)$. The logistic map hints that some analysis of the steady-state ought to be possible, certainly for $1 < r < 3$ where a single solution is predicted. In this case:

$$x = rx(1-x) \therefore \tfrac{1}{r}x = x - x^2$$
$$\Rightarrow x \left(x - 1 + \tfrac{1}{r} \right) = 0$$
$$\Rightarrow x = 0, \; x = 1 - \tfrac{1}{r} \tag{4.4}$$

This means steady-state values of $x = 0$ when $r = 1$; $x = \frac{1}{2}$ when $r = 2$, and $x = \frac{2}{3}$ when $r = 3$. These are consistent with the numerical results of the logistic map. However, the logistic map shows that this analysis is inappropriate for $3 < r < 4$, since the x values don't settle on one particular value.

Now for an iteration of the form $x_{n+1} = f(x_n)$ to terminate at a fixed point $x = x^*$, one can clearly see by drawing a 'cobweb diagram' of $y = x$ and $y = f(x)$ that $\left|\frac{df}{dx}\right|_{x^*} < 1$. i.e. the magnitude of the gradient of $y = f(x)$ must be less than the gradient of $y = x$ (i.e. a gradient limit of unity).

For $1 \le r < 3$, $f(x) = rx(1 - x)$ $\therefore \frac{df}{dx} = r - 2rx$. Hence if $\left|\frac{df}{dx}\right|_{x^*} < 1$ and $x^* = 1 - \frac{1}{r}$, this means $\left|r - 2r\left(1 - \frac{1}{r}\right)\right| < 1$ and $\therefore |2 - r| < 1 \Rightarrow 1 < r < 3$. This is consistent with Fig. 4.2. So in summary, the curve $x(r) = 1 - \frac{1}{r}$ describes the asymptotic value of the logistic map between $r = 1$ and $r = 3$.

4.3.5 *Describing $x(r)$ for the first bifurcation*

In the previous section we demonstrated $x(r) = 1 - \frac{1}{r}$ is the asymptotic behaviour within range $1 \le r < 3$. Between $r = 3$ and about $k = 3.45$, the asymptotic population fraction $x(r)$ bifurcates, meaning the long term population tends to oscillate between two values. Can we determine the functional form of $x(r)$ within this region, and indeed determine an exact value for k, i.e. the r value where a further cascade of bifurcations begins?

The oscillatory behaviour between the first bifurcation at $r = 3$ and the second bifurcation at $r = k$ can be used to find k. From Fig. 4.2, it appears to be about 3.45. In the range $3 \le r \le k$, the asymptotic x values repeat every *second* iteration. In other words:

$$x_1 = rx(1 - x), \quad x = rx_1(1 - x_1) \therefore x = r\left\{rx(1 - x)\right\}\left(1 - \left\{rx(1 - x)\right\}\right) \quad (4.5)$$

Hence:

$$x = r\left(rx - rx^2\right)\left(1 - rx + rx^2\right)$$

i.e.

$$x\left\{r^2x^3 - 2r^2x^2 + (r + 1)rx + \frac{1}{r} - r\right\} = 0 \quad (4.6)$$

Now consider the cubic:

$$P(x) = \left(x - 1 + \frac{1}{r}\right)\left(r^2x^2 - \left(r^2 + r\right)x + r + 1\right) \quad (4.7)$$

Multiplying out:

$$P(x) = x^3(r^2) + x^2\left(-r^2 + r - r^2 - r\right) + x\left(r + 1 - (r^2 + 4)\left(\frac{1}{r} - 1\right)\right)$$

$$+ (r+1)\left(\frac{1}{r} - 1\right)$$

$$P(x) = r^3 x^3 - 2r^2 x^2 + x\left(r + 1 - \left\{r + 1 - r^2 - r\right\}\right) + 1 + \frac{1}{r} - r - 1$$

$$P(x) = r^2 x^3 - 2r^2 x^2 + (r+1)rx + \frac{1}{r} - r \tag{4.8}$$

So since $x\left\{r^2 x^3 - 2r^2 x^2 + (r+1)rx + \frac{1}{r} - r\right\} = 0$, and $x \neq 0$ in general when $3 \leq r < 4$, this means $r^2 x^3 - 2r^2 x^2 + (r+1)rx + \frac{1}{r} - r = P(x) = 0$. Now when $r = 3$, $x = 1 - \frac{1}{r}$, which means $x - 1 + \frac{1}{r} = 0$. Hence the $x_\pm$ values at the first 'pitchfork' bifurcation are therefore the solutions to the quadratic factor of $P(x)$:

$$r^2 x^2 - \left(r^2 + r\right)x + r + 1 = 0 \tag{4.9}$$

Using the quadratic formula:

$$x = \frac{r^2 + r \pm \sqrt{\left(r^2 + r\right)^2 - 4r^2\left(r + 1\right)}}{2r^2}$$

$$x = \frac{r + 1 \pm \frac{1}{r}\sqrt{r^2\left(r + 1\right)^2 - 4r^2\left(r + 1\right)}}{2r}$$

$$x = \frac{r + 1 \pm \sqrt{\left(r + 1 - 4\right)\left(r + 1\right)}}{2r}$$

$$\therefore x_\pm = \frac{r + 1 \pm \sqrt{(r - 3)(r + 1)}}{2r} \tag{4.10}$$

For the first bifurcation region of $x_{n+1} = f(x_n)$, i.e. the asymptotic behaviour of x_∞ within region $3 \leq r \leq k$, the analysis above gives:

$$f(x) = r(rx - rx^2)(1 - rx + rx^2) \tag{4.11}$$

Differentiating this and then substituting $x_\pm = \frac{r+1\pm\sqrt{(r-3)(r+1)}}{2r}$ into our stability criteria $\left|\frac{df}{dx}\right|_{x^*} < 1$ seems rather tedious. Instead let us deploy a clever trick that Strogatz uses (p. 360). Firstly note, $f(x) = gg(x) = g(z(x))$ where $z(x) = rx(1 - x) = rx - rx^2$ and $g(z) = rz - rz^2$. i.e. in our first bifurcation region, we apply the logistic map *twice* to get back to the same x value, asymptotically. Therefore, using the chain rule:

$$\frac{df}{dx} = \frac{dg}{dz} \times \frac{dz}{dx}$$

$$\frac{df}{dx} = (r - 2rz)(r - 2rx) \tag{4.12}$$

Now in our first bifurcation, if $x = x_-$, the next iteration $z = g(x_-)$, must be x_+. Therefore, our bifurcation stability criteria $\left| \frac{df}{dx} \right|_{x^*} < 1$ implies:

$$|(r - 2rx_+)(r - 2rx_-)| < 1$$

$$\left| r^2 - 2r^2(x_+ + x_-) + 4r^2 x_+ x_- \right| < 1 \tag{4.13}$$

Now:

$$x_+ + x_- = \frac{r + 1 + \sqrt{(r-3)(r+1)}}{2r} + \frac{r + 1 - \sqrt{(r-3)(r+1)}}{2r} = \frac{r + 1}{r}$$

$$x_+ x_- = \frac{1}{4r^2} \left(r + 1 + \sqrt{(r-3)(r+1)} \right) \left(r + 1 - \sqrt{(r-3)(r+1)} \right)$$

$$\therefore x_+ x_- = \frac{1}{4r^2} \left((r+1)^2 - (r-3)(r+1) \right)$$

$$= \frac{(r+1)}{4r^2} (r + 1 - r + 3) = \frac{(r+1)}{r^2} \tag{4.14}$$

Therefore,

$$\left| r^2 - 2r^2(x_+ + x_-) + 4r^2 x_+ x_- \right| < 1$$

$$\left| r^2 - 2r(r + 1) + 4(r + 1) \right| < 1$$

$$\left| -r^2 + 2r + 4 \right| < 1$$

$$\left| -(r - 1)^2 + 5 \right| < 1 \tag{4.15}$$

Hence for stability of the single bifurcation $\left| -(r-1)^2 + 5 \right| < 1$, i.e. $r_- \leq r < r_+$. To find the limits of this range, consider the solutions of:

$$-(r - 1)^2 + 5 = 1$$

$$4 = (r - 1)^2$$

$$1 \pm 2 = r \tag{4.16}$$

Since $r > 0$ this means $r = 3$ is the lower bound r_- for r in $\left| -(r-1)^2 + 5 \right| < 1$. For the upper limit r_+ consider:

$$(r - 1)^2 - 5 = 1$$

$$(r - 1)^2 = 6$$

$$r = 1 \pm \sqrt{6} \tag{4.17}$$

i.e. since $r > 1$ $\therefore r = 3$ or $1 + \sqrt{6}$. Therefore, the next bifurcation should occur at $r = 1 + \sqrt{6} \approx 3.45$, as suggested previously.

In summary for $x_{n+1} = rx_n(1 - x_n)$:

- $0 < r < 1 : x_\infty \to 0$
- $1 \leq r < 3 : x_\infty \to 1 - \frac{1}{r}$
- $3 \leq r < 1 \pm \sqrt{6} : x_\infty \to \frac{r+1\pm\sqrt{(r-3)(r+1)}}{2r}$ i.e. a 'pitchfork bifurcation'
- $1 \pm \sqrt{6} \leq r < 4 :$ Progressive cascade of bifurcations, with successive bifurcation intervals tending towards the Feigenbaum number ratio:

$$\lim_{n \to \infty} \left(\frac{r_n - r_{n-1}}{r_{n+1} - r_n} \right) = 4.669201609\ldots$$

4.3.6 *The logistic map is a fractal*

If one sets up a computer program to enable r_{min} and r_{max} to be varied and the logistic map to be automatically recalculated, it is possible to 'zoom in' to a region of the r, x space. Doing this (see Fig. 4.3) reveals another characteristic feature of many chaotic systems, that of a *self-similar geometry*. This means that as one zooms to higher magnification, the structure repeats itself. A fixed pattern repeats *at all scales*. This is a very common feature of natural systems, from the mountain-like rocks which have mountain-like electron-micrographs, to the eddies within eddies that characterize turbulent fluid flow. The logistic map is infinitely complex, yet it possesses a recursive symmetry, a repeating signature motif. Benoit Mandelbrot (1924–2010) coined the term *fractal* to describe self-similar geometries. Fractal refers to a 'fractional dimension' metric different scales of rulers (1D) or boxes (2D). Gleick's *Chaos* [17] describes this very nicely, and his superb book motivated me to create some simulations myself. These are illustrated in a lecture *A Computational Cookbook of Chaos* that can be accessed from http://www.eclecticon.info/lectures3.htm.

In Fig. 4.3, zooming in reveals the infinite tree of bifurcations, and enables a mechanism for confirming Feigenbaum's hypothesis. Figure 4.3(b) shows that as the May system proceeds towards full-blown chaos when $r \to 4$, the probability of any given value of x tends to be *uniformly distributed*, i.e. is a fixed colour in the plot of the probability of a given x value plotted against r. If all values of x in the range [0,1] and uniformly distributed, this is another way of saying the x values are effectively random. The probabilities are $1, \frac{1}{2}, \frac{1}{4}, \frac{1}{8}$ etc for the cascade of bifurcations, i.e. asymptotic cycles between 2^n values, where n is an integer. Intriguingly, Strogatz reports that a period-3 'trifurcation' (i.e. a three way split) occurs at $3.8284 \leq r \leq 3.8415$, which will of course change the probability distributions somewhat. Furthermore, (a) shows there are 'stripes of stability' in all this chaos, and these too have a self-similar, fractal geometry.

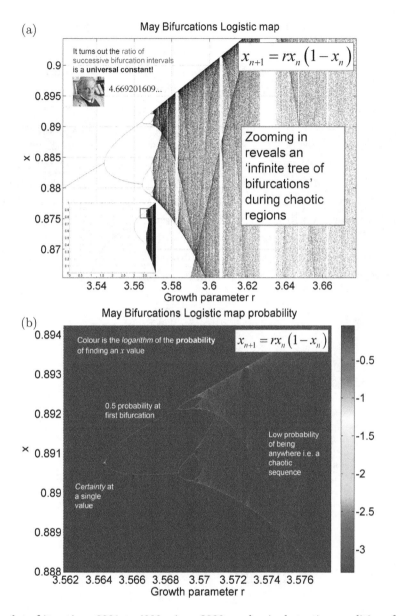

Fig. 4.3 A plot of iterations 2001 to 4000, given 2000 randomised starting conditions for x within [0,1], against growth parameter r of May's logistic equation. Zooming in (and recalculating, given a fresh thousand samples from a more restricted x interval) reveals the infinite tree of bifurcations, and enables a mechanism for confirming Feigenbaum's hypothesis. (b) shows that as the May system proceeds towards full-blown chaos when $r \to 4$, the probability of any given value of x tends towards a uniform distribution, i.e. is represented by a fixed colour in the plot of the probability of a given x value plotted against r. If all values of x in the range [0,1] are uniformly distributed, this is another way of saying the x values are effectively random. For the curves describing the cascade of bifurcations, the probabilities are 1, $\frac{1}{2}$, $\frac{1}{4}$, $\frac{1}{8}$, etc. Note that period-3 'trifurcations' (i.e. a three-way split) are also possible, which will of course change the probability distributions somewhat. Furthermore, (a) shows there are 'stripes of stability' in all this chaos, and these too have a self-similar, *fractal* geometry.

4.4 The Need for Numerical Methods and Nonlinearity

The simulation of May's bifurcations is a classic gateway to Chaos theory ideas, but I think it offers much more to general mathematics education. A comprehensive understanding of quadratic equations is absolutely foundational at pre-University study, and May's logistic map is a wonderfully rich context, complete with many numerical surprises in the chaotic regime of $r \to 4$. The 2020 pandemic has forced widespread remote learning upon millions of students, and as a consequence, access to education means access to a computer. This could be a golden opportunity to incorporate numerical, and programmatically graphical, methods into basic mathematics teaching, and allow the 'non-elephants' of non-linear equations to be explored using spreadsheets or basic coding environments. It will also enable wider information-technology skills to be developed in normal (rather than specialised) lessons, skills that will prove essential in a world of employment and problem solving that exist beyond the constraints of schools and examinations.

Chapter 5

Pendulums, Poincaré Diagrams and Strange Attractors

5.1 A Phase Portrait of the Possible Trajectories of a Dynamical System

In the previous chapter, *May's Chaotic Bunnies*, the relationship of fractional rabbit population x_n with year number n becomes increasingly *chaotic* as the growth parameter r increases from 3 to 4. Although the underpinning equation of the system $x_{n+1} = rx_n(1 - x_n)$ is both simple and entirely deterministic, highly complex behaviour can nonetheless manifest. The *logistic map* idea proved useful in making some sense of the apparent randomness in x vs n 'trajectories', as r tended to 4. Rather than attempting to characterise x_n given a particular starting value x_0, *all* possible trajectories are overlaid on the same graph, given random starting values. In this way, Figs. 4.2 and 4.3 reveal the major patterns. As r varies, we move from extinction to a single asymptotic value, a pitchfork bifurcation (i.e. an oscillation between two fixed values), then a cascade of bifurcations (meaning 4,8,16 cycles), plus some 3-cycle oddities too.

Henri Poincaré (1854–1912) bequeathed many mathematical gifts, and one of these is the idea of a *phase portrait* (which I will often refer to as a Poincaré diagram), to graphically represent all possible solutions to a differential equation associated with a dynamical system, i.e. one that varies with time. The key idea is to not plot the time variation directly, but use it in a *parametric* sense. Before we tackle some chaotic systems (a double-pendulum, and then the Lorenz and Rössler *strange attractors*), let us illustrate the Poincaré diagram idea via a re-imagining of the dynamics of a simple pendulum.

5.2 Poincaré's Pendulum

Consider an idealised simple pendulum (see Fig. 5.1) comprising a small mass m attached to a light, inextensible string of length l to a frictionless pivot. Let two forces act on the pendulum mass. The first is gravity, with strength mg acting vertically downwards, and air resistance, which will always oppose motion. If one considers air slamming into the pendulum mass as it oscillates, the rate of change of momentum of the air will vary as v^2 where v is the pendulum speed. This is because the mass of air swept per second is proportional to v and therefore the momentum exchanged per second must vary as v^2. If the pendulum string is always taut, then $v = l\dot{\theta}$, where θ is the angle of the pendulum string from the vertical measured in radians. The symbol $\dot{\theta} = \frac{d\theta}{dt}$, i.e. the rate of change of angle, or 'angular velocity'.

Newton's second law (mass $\times$ acceleration = vector sum of force), evaluated in a tangential direction to the circular path of the pendulum mass is:

$$ml\frac{d\dot{\theta}}{dt} = -mg\sin\theta - kl^2\dot{\theta}\left|\dot{\theta}\right| \tag{5.1}$$

Note by expressing the air resistance force as $-kl^2\dot{\theta}\left|\dot{\theta}\right|$, we incorporate the v^2 proportionality in its magnitude, but also guarantee that it always opposes motion, whose direction is characterised by the sign of θ.

If we can ignore air resistance, and the pendulum angular deviation θ is small such that $\sin\theta \approx \theta$:

$$\frac{d\dot{\theta}}{dt} \approx -\frac{g}{l}\theta \tag{5.2}$$

i.e. *Simple Harmonic Motion* (SHM) with solution:

$$\theta(t) = \theta_0\cos\left(\frac{2\pi t}{P}\right) \tag{5.3}$$

$$P = 2\pi\sqrt{\frac{l}{g}} \tag{5.4}$$

i.e. an oscillatory time variation with period P. In order to tease out general behaviour, define a dimensionless timescale $t = P\tau$ and re-write $\dot{\theta} \to \frac{1}{P}\dot{\theta}$ where $\dot{\theta} = \frac{d\theta}{d\tau}$.

Hence:

$$\frac{ml}{P^2}\frac{d\dot{\theta}}{d\tau} = -mg\sin\theta - \frac{kl^2}{P^2}\dot{\theta}\left|\dot{\theta}\right|$$

$$\frac{d\dot{\theta}}{d\tau} = -\frac{g}{l}P^2\sin\theta - \frac{kl}{m}\dot{\theta}\left|\dot{\theta}\right|$$

$$\frac{d\dot{\theta}}{d\tau} = -4\pi^2\sin\theta - \frac{kl}{m}\dot{\theta}\left|\dot{\theta}\right| \tag{5.5}$$

If the pendulum bob is spherical, with radius r, and it swings through air of density ρ:

$$k = \tfrac{1}{2}c_D\rho\pi r^2 \tag{5.6}$$

where c_D is the drag coefficient. (Between 0.1 and 0.5 might be typical values [30, pp. 148–156]).

Define a dimensionless constant a:

$$a = \frac{kl}{m} = \frac{\frac{1}{2}c_D\rho\pi r^2 l}{m} \tag{5.7}$$

i.e. a is $\frac{1}{2}c_D$ multiplied by the ratio of the mass of air equivalent to the pendulum moving through a cylinder of length l, to the actual mass of the pendulum.

Unfortunately, $\frac{d\dot\theta}{d\tau} = -4\pi^2 \sin\theta - a\dot\theta\left|\dot\theta\right|$ does not have an analytical solution. Another example of nonlinearity being the cause of trouble! However, we can calculate the (approximate) dynamics nonetheless by using a numerical solution scheme. As discussed in *Numerical methods as an introduction to calculus* [12], an Euler method is exponentially error prone for SHM, and therefore the next simplest scheme is perhaps the *Velocity Dependent Acceleration Verlet* (VDAV) method, i.e. constant acceleration motion between finite time steps $\Delta\tau$, for evaluating the pendulum dynamics,[1] with initial conditions: $\tau = 0$, $\theta = \theta_0$, $\dot\theta = 0$. Note in VDAV we do actually have to use an Euler-method to work out a correction $\ddot\phi$ to the angular acceleration, which sidesteps having to use the updated angular velocity $\dot\theta_{n+1}$ that we are trying to calculate, i.e. we avoid a *circular reference*, which is rather ironic given the geometry of the system!

The constant (angular) acceleration Verlet iterative step is:

$$\ddot\theta_n = -4\pi^2 \sin\theta_n - a\dot\theta_n\left|\dot\theta_n\right| \tag{5.8}$$

$$\theta_{n+1} = \theta_n + \dot\theta_n\Delta\tau + \tfrac{1}{2}\ddot\theta_n(\Delta\tau)^2$$

The angular velocity can then be updated, using the Euler-method correction, by:

$$\dot\phi = \dot\theta_n + \ddot\theta_n\Delta\tau$$

$$\ddot\phi = -4\pi^2 \sin\theta_{n+1} - a\dot\phi\left|\dot\phi\right|$$

$$\dot\theta_{n+1} = \dot\theta_n + \tfrac{1}{2}\left(\ddot\theta_n + \ddot\phi\right)\Delta\tau$$

The results of the numerical scheme above are evaluated for $0 \le \tau \le 10$, and $\Delta\tau = 0.01$, and plotted in Fig. 5.1. An 'exact' SHM solution $\theta(\tau) = \theta_0\cos(2\pi\tau)$, i.e. when $a = 0$ and $\theta \ll 1$, is compared to a large amplitude, air-resistance-free trace (i.e. $a = 0$, but θ not always small). The same starting condition ($\theta = 0.5$) as the second trace, but with $a = 1$, is also plotted. The second larger amplitude trace oscillates in a very regular fashion as the SHM solution, but the period is slightly

---

[1]A more efficient method is the *Runge–Kutta* scheme, which has errors which scale as Δt^4, whereas *Verlet* errors scale as Δt^2. See Appendix B for a description of the Runge–Kutta method, applied to this pendulum situation.

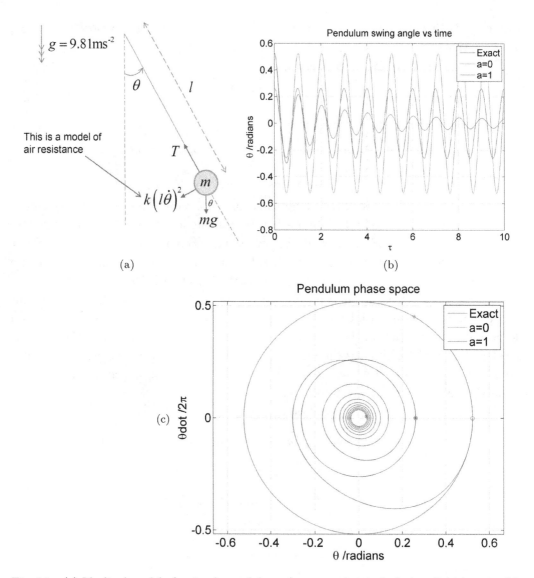

Fig. 5.1   (a) Idealised model of a simple pendulum. A mass m is attached via a light inextensible string of length l to a frictionless pivot, and released from rest at angle θ_0 to the vertical. For small angles, and if air resistance can be ignored, the pendulum undergoes SHM with period $P = 2\pi\sqrt{\frac{l}{g}}$. If $\theta \ll 1$ is not a valid approximation then we must endeavour to solve the equation of motion (from Newton's Second Law) $ml\frac{d\dot{\theta}}{dt} = -mg\sin\theta - kl^2\dot{\theta}\left|\dot{\theta}\right|$ using a numerical method such as *Verlet* (constant acceleration motion between fixed time steps) or the even more precise *Runge–Kutta* method (see Appendix B). (b) compares the exact SHM solution (blue) to an equivalent larger initial angle version. The green trace is oscillatory, but goes out of phase with the exact solution. The period *does depend on amplitude for large angles*. The red trace represents the same initial parameters as the green trace, but also includes the effect of air resistance. This causes a decay of amplitude, and an inspiralling in the Poincaré diagram (or 'phase portrait') in (c), formed by plotting $\frac{1}{2\pi}\dot{\theta}$ vs θ.

longer and as a result, it goes out of phase within the $\tau = [0, 10]$ interval. The $a = 1$ trajectory decays, but interestingly, is closer in phase with the SHM solution. This is probably to be expected as θ reduces, since the nonlinear $\sin \theta$ terms become closer to the linear θ.

However, the most interesting, and in my opinion, aesthetic representation, is the Poincaré diagram. In this case $y = \frac{1}{2\pi}\dot{\theta}$ is plotted against $x = \theta$ for each of the three scenarios. For the exact SHM solution, this means $y = -\theta_0 \sin(2\pi\tau)$ and $x = \theta_0 \cos(2\pi\tau)$, which is the equation of a (blue) circle of radius θ_0. As scaled time τ progresses, the x, y trajectory of the pendulum in the Poincaré diagram is that of a clockwise progression along a circle with Cartesian equation $x^2 + y^2 = \theta_0^2$. The approximately fixed period (well on the timescale shown) of the large amplitude pendulum is indicated by the larger (green) circle. The $a = 1$ solution is represented by an inspiralling curve, showing the extent that both oscillation amplitude and pendulum speed simultaneously diminish with time due to the energy-sapping impact of air resistance.

5.3 The Double Pendulum

Let us use the ideas developed for the simple pendulum to consider a *coupled system*. A simple pendulum with string length l_1 and bob mass m_1 is itself attached to another simple pendulum of string length l_2 and bob mass m_2. Clearly the 'light inextensible string' approximation might be better represented by rigid rods here, and the upshot is a slight reduction of l due to a change in position of centre of mass, and the incorporation of the rod moments of inertia in the equivalent expression for Newton's second law. However, the idea is to demonstrate a mathematical paradigm rather than construct a realistic physical model here, so I am going to continue with the simple pendulum concept. I am also going to ignore air resistance, pivot friction and any other losses. As you will see, the system becomes quite complex without any of these additions!

Referring to Fig. 5.2, the x, y coordinates of pendulum masses are:

$$x_1 = l_1 \sin \theta_1 \tag{5.9}$$

$$y_1 = -l_1 \cos \theta_1 \tag{5.10}$$

$$x_2 = l_1 \sin \theta_1 + l_2 \sin \theta_2 \tag{5.11}$$

$$y_2 = -l_1 \cos \theta_1 - l_2 \cos \theta_2 \tag{5.12}$$

The corresponding x, y velocities of pendulum masses are:

$$v_{x1} = l_1 \cos \theta_1 \dot{\theta}_1 \tag{5.13}$$

$$v_{y1} = l_1 \sin \theta_1 \dot{\theta}_1 \tag{5.14}$$

$$v_{x1} = l_1 \cos \theta_1 \dot{\theta}_1 + l_2 \cos \theta_2 \dot{\theta}_2 \tag{5.15}$$

$$v_{y2} = l_1 \sin \theta_1 \dot{\theta}_1 + l_2 \sin \theta_2 \dot{\theta}_2 \tag{5.16}$$

The double pendulum

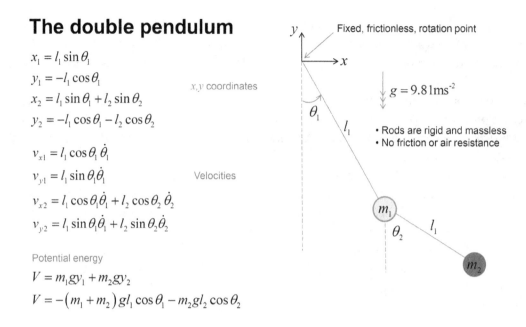

$$x_1 = l_1 \sin\theta_1$$

$$y_1 = -l_1 \cos\theta_1$$

x,y coordinates

$$x_2 = l_1 \sin\theta_1 + l_2 \sin\theta_2$$

$$y_2 = -l_1 \cos\theta_1 - l_2 \cos\theta_2$$

$$v_{x1} = l_1 \cos\theta_1\,\dot\theta_1$$

$$v_{y1} = l_1 \sin\theta_1\dot\theta_1$$

Velocities

$$v_{x2} = l_1 \cos\theta_1\dot\theta_1 + l_2 \cos\theta_2\,\dot\theta_2$$

$$v_{y2} = l_1 \sin\theta_1\dot\theta_1 + l_2 \sin\theta_2\dot\theta_2$$

Potential energy

$$V = m_1 g y_1 + m_2 g y_2$$

$$V = -\left(m_1 + m_2\right)gl_1 \cos\theta_1 - m_2 g l_2 \cos\theta_2$$

Kinetic energy

$$T = \tfrac{1}{2}m_1\left(v_{x1}^2 + v_{y1}^2\right) + \tfrac{1}{2}m_2\left(v_{x2}^2 + v_{y2}^2\right)$$

$$T = \tfrac{1}{2}m_1 l_1^2\dot\theta_1^2 + \tfrac{1}{2}m_2\left[l_1^2\dot\theta_1^2 + l_2^2\dot\theta_2^2 + 2l_1 l_2\dot\theta_1\dot\theta_2 \cos\left(\theta_1 - \theta_2\right)\right]$$

Fig. 5.2 A model of an idealised coupled pendulum. Masses m_1 and m_2 are connected by light and rigid rods of length l_1 and l_2 to frictionless rotating joints. Air resistance is also ignored, so the sum of kinetic and potential energy (the latter purely due to gravity of strength g) is constant. The system is significantly more complex than a simple pendulum, and therefore an opportunity to demonstrate the power of the *Lagrangian* method of generating the equations of motion for angles θ_1 and θ_2, which can then be solved via an appropriate numerical method such as *Verlet* or *Runge–Kutta*.

The total potential energy (in this case, solely due to work done against gravity) is:

$$V = m_1 g y_1 + m_2 g y_2 \tag{5.17}$$

$$V = -(m_1 + m_2)g l_1 \cos\theta_1 - m_2 g l_2 \cos\theta_2 \tag{5.18}$$

And the total kinetic energy of the system is:

$$T = \tfrac{1}{2}m_1\left(v_{x1}^2 + v_{y1}^2\right) + \tfrac{1}{2}m_2\left(v_{x2}^2 + v_{y2}^2\right)$$

$$T = \tfrac{1}{2}\left(m_1 + m_2\right)l_1^2\dot\theta_1^2 + \tfrac{1}{2}m_2 l_2^2\dot\theta_2^2 + m_2 l_1 l_2\dot\theta_1\dot\theta_2 \cos\left(\theta_1 - \theta_2\right) \tag{5.19}$$

How to solve the dynamics of a system of more than one variable can be tricky using Newton's laws. In this case, the pair of variables are the angles θ_1 and θ_2. An elegant, and systematic recipe, is to determine the *Lagrangian* $L = T - V$ and to solve the *Euler-Lagrange (EL) equations*.[2] All the required ingredients are the *scalar equations* for kinetic and potential energy. There will be one EL equation per variable, which form the set of differential equations that need to be solved (numerically) in order to construct a simulation of the double pendulum. The Lagrangian L for the double-pendulum is:

$$L = \underbrace{\tfrac{1}{2}\left(m_1 + m_2\right) l_1^2 \dot{\theta}_1^2 + \tfrac{1}{2} m_2 l_2^2 \dot{\theta}_2^2 + 2 l_1 l_2 \dot{\theta}_1 \dot{\theta}_2 \cos\left(\theta_1 - \theta_2\right)}_{\text{Kinetic energy } T}$$

$$+ \underbrace{(m_1 + m_2) g l_1 \cos\theta_1 + m_2 g l_2 \cos\theta_2}_{\text{Potential energy } V}$$

and the Euler–Lagrange equations are:

$$\frac{d}{dt}\left(\frac{\partial L}{\partial \dot{\theta}_1}\right) = \frac{\partial L}{\partial \theta_1} \tag{5.20}$$

$$\frac{d}{dt}\left(\frac{\partial L}{\partial \dot{\theta}_2}\right) = \frac{\partial L}{\partial \theta_2} \tag{5.21}$$

The leads[3] to the following *Verlet* scheme with fixed timestep $\Delta t = \frac{1}{1000} 2\pi \sqrt{\frac{l_1 + l_2}{g}}$ with $\theta = \theta_1$, $\phi = \theta_2$, and $\Delta_n = \phi_n - \theta_n$ with initial conditions $\tau = 0$, $\theta = \theta_0$, $\dot{\theta} = 0$, $\phi = \phi_0$, $\dot{\phi} = 0$:

$$\ddot{\theta}_n = \frac{m_2 l_1 \dot{\theta}_n^2 \sin\Delta_n \cos\Delta_n + m_2 g \sin\phi_n \cos\Delta_n + m_2 l_2 \dot{\phi}_n^2 - (m_1 + m_2) g \sin\theta_n}{(m_1 + m_2) l_1 - m_2 l_1 \cos^2\Delta_n}$$

$$\ddot{\phi}_n = \frac{-m_2 l_2 \dot{\phi}_n^2 \sin\Delta_n \cos\Delta_n + (m_1 + m_2)\left(g\sin\theta_n \cos\Delta_n - l_1 \dot{\theta}_n^2 \sin\Delta_n - g\sin\phi_n\right)}{(m_1 + m_2) l_2 - m_2 l_2 \cos^2\Delta_n}$$

$$\theta_{n+1} = \theta_n + \dot{\theta}_n \Delta t + \tfrac{1}{2}\ddot{\theta}_n(\Delta t)^2$$

$$\phi_{n+1} = \phi_n + \dot{\phi}_n \Delta t + \tfrac{1}{2}\ddot{\phi}_n(\Delta t)^2$$

$$\omega = \dot{\theta}_n + \ddot{\theta}_n \Delta t$$

$$\Omega = \dot{\phi}_n + \ddot{\phi}_n \Delta t \tag{5.22}$$

---

[2]For those new to the idea of the *Calculus of Variations* it may seem like I have waved a mathematical magic wand here. And you would be correct. In some ways, the Euler–Lagrange equations *are* like a magic wand. I have written notes on their derivation and use which can be accessed from: http://www.eclecticon.info/index_htm_files/Calculus%20-%20Variational %20Calculus.pdf and http://www.eclecticon.info/index_htm_files/Mechanics%20-%20Lagrangians. pdf.

[3]I have skipped quite a bit of algebra here! The goal is to use the Euler–Lagrange equations to determine equations for the angular accelerations $\ddot{\theta}$ and $\ddot{\phi}$. I took my starting point from: http:// scienceworld.wolfram.com/physics/DoublePendulum.html.

$$\dot{\omega} = \frac{m_2 l_1 \omega^2 \sin \Delta_{n+1} \cos \Delta_{n+1} + m_2 g \sin \phi_{n+1} \cos \Delta_{n+1} + m_2 l_2 \Omega^2 - (m_1 + m_2) g \sin \theta_{n+1}}{(m_1 + m_2) l_1 - m_2 l_1 \cos^2 \Delta_{n+1}}$$

$$\dot{\Omega} = \frac{m_2 l_2 \Omega^2 \sin \Delta_{n+1} \cos \Delta_{n+1} + (m_1 + m_2)(g \sin \theta_{n+1} \cos \Delta_{n+1} - l_1 \omega^2 \sin \Delta_{n+1} - g \sin \phi_{n+1})}{(m_1 + m_2) l_2 - m_2 l_2 \cos^2 \Delta_{n+1}}$$

$$\dot{\theta}_{n+1} = \dot{\theta}_n + \tfrac{1}{2}\left(\dot{\theta}_n + \dot{\omega}\right) \Delta t$$

$$\dot{\phi}_{n+1} = \dot{\phi}_n + \tfrac{1}{2}\left(\dot{\phi}_n + \dot{\Omega}\right) \Delta t \tag{5.23}$$

A more precise *Runge–Kutta* scheme is provided in Appendix B. Figures 5.3–5.5, compare the regular motion of $m_1 = 1, m_2 = 3, l_1 = 3, l_2 = 2$ (in kg, m respectively),

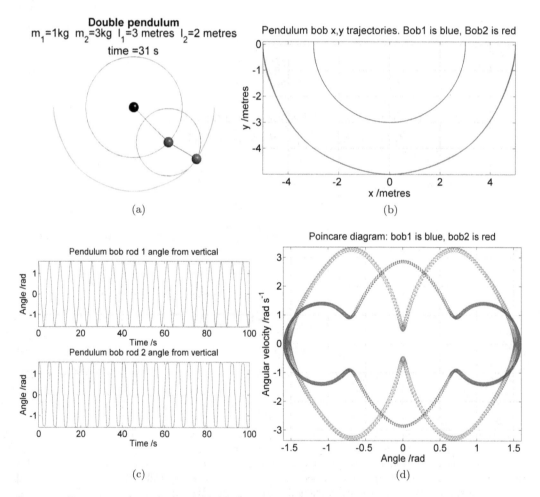

Fig. 5.3   Dynamics of an idealised double (i.e. coupled) pendulum, as illustrated in subplot (a), a frame from a video sequence generated from a MATLAB implementation of the equations of motion developed in this chapter. In this case, motion is very regular, as indicated by the x, y trajectories of the pendulum bobs in (b), the co-sinusoidal time variations in (c), and the phase portrait in (d).

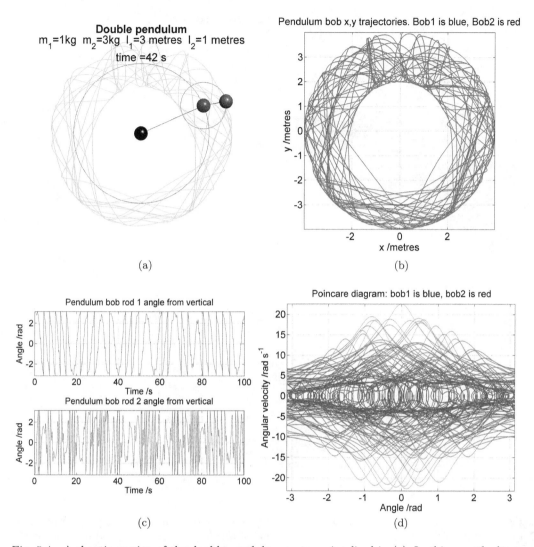

Fig. 5.4   A chaotic version of the double pendulum system visualised in (a). In this case the lower mass length l_2 is reduced to 1 m, and the system is released from an angle of $\theta_0 = 0.99\pi$. The trajectories are somewhat chaotic, as indicated by the time variation in (c). The x, y trajectories in (b) and the phase portrait in (d) indicate a significantly larger set of the possible $\theta, \dot{\theta}$ phase space is traversed than in the non-chaotic version. However, the phase portrait is not without structure and is clearly symmetric in θ. The motion is complex, but *not* random.

released from $\theta_0 = \frac{\pi}{2}, \phi_0 = 0$, with the wild chaos of $m_1 = 1, m_2 = 3, l_1 = 3, l_2 = 1$, released from $\theta_0 = 0.99\pi$ and $\phi_0 = 0$. The strength of gravity is $g = 9.81\,\mathrm{Nkg}^{-1}$. Although the motions are best seen as videos (click on the red triangle hyperlinks below the *Computational Cookbook of Chaos* presentation in http://www.eclecticon.info/lectures3.htm), Figs. 5.3 and 5.4 contrast the two scenarios. In the latter chaotic scenario, both pendulum bobs appear to occupy almost every possible

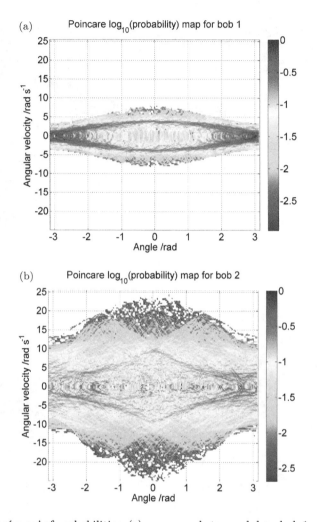

Fig. 5.5   A 'Poincaré map' of probabilities. (a) corresponds to pendulum bob 1 and (b) corresponds to pendulum bob 2. To construct the coloured plots of this figure, I divided a grid based upon the maximum and minimum angle and angular speed values into 200 'bins' for each dimension. I then counted how many trajectories past through each of these 200×200 grid squares. These numbers were then scaled by the total number of trajectory points, which was $N = 100,001$, and then $\log_{10}$ of these scaled frequencies was mapped to the blue to red colour scale. This illustrates the major differences between pendulum bob 1 and bob 2, in the chaotic scenario. Bob 1 is confined to largely elliptical trajectories in the $\theta, \dot{\theta}$ space, with occasional departures, whereas bob 2 occupies much more of the 'phase space'. However, the probability map has definite structure: it has symmetry and characteristic patterns. The motion of bob 2 may be chaotic, but its motion in $\phi, \dot{\phi}$ phase space is far from uniformly random.

angle, although the phase portrait implies there are combinations of angle and angular velocity that are either not allowed due to conservation of energy constraints, or simply very unlikely. To better investigate the latter, I have constructed a Poincaré map of *probabilities*. To construct the coloured plots of Fig. 5.5, I divided a grid based upon the maximum and minimum angle and angular speed values into 200 'bins' for each respective 'dimension' of the Poincaré map. I then counted how many trajectories past through each of these 200×200 grid squares. These numbers were then scaled by the total number of trajectory points, which was $N = 100,001$, and then $\log_{10}$ was taken of this scaled frequency, to relate to the blue to red colour scale. This process illustrates the major differences between pendulum bob 1 and bob 2, in the chaotic scenario. Bob 1 is confined to largely elliptical trajectories in the $\theta, \dot{\theta}$ space, with occasional departures, whereas bob 2 occupies much more of the *phase space* of the Poincaré map. Overall, the probability map has definite structure: it has symmetry, and characteristics patterns. The motion of bob 2 may be chaotic, but its motion in $\phi, \dot{\phi}$ phase space is far from uniformly random.

5.4 Lorenz and Rössler Strange Attractors

The story goes that Edward Lorenz (1917–2008) was using a Royal McBee LGP-30 computer in 1961 to model weather patterns. To print in those days was literally to produce a new line of numbers via some form of dot-matrix printer. For those within a certain age bracket, the unique rasping sound of a shuttling print head needling inked ribbon, will immediately be recalled from memories of a distant, and noisier, office past. Lorenz decided to save time and commence his simulation partway through where he had got to the previous day. Unfortunately in doing this, he inadvertently introduced a small rounding error. The print-outs were numbers to three significant figures, whereas the computer actually worked to six-digit precision. You might think such a tiny error would have a very small effect, but Lorenz noticed that his equations were so sensitive to initial conditions, that his model eventually produced completely different weather, compared to what was predicted at the end of the previous day, *without* his manual short-cut. Like any circumspect scientist, Lorenz would have checked both scenarios. Lorenz's equations were behaving like a pool break; cueing the white ball with a tiny difference in impulse produced a totally different final position of the coloured balls.

Lorenz's system of equations (e.g. [49, pp. 311–334, 45, p. 129]) were something like:

$$\frac{dx}{dt} = s(y - x)$$

$$\frac{dy}{dt} = x(r - z) - y$$

$$\frac{dz}{dt} = xy - bz \tag{5.24}$$

with parameters $s = 10$, $r = 28$, $b = \frac{8}{3}$. I have chosen the variables x, y, z deliberately, as the next step is to overlay all possible trajectories in a Cartesian x, y, z phase space, rather than focus on a time-series analysis, just as we did for Poincaré's single and double pendulums. The result is remarkable. Although the time variation

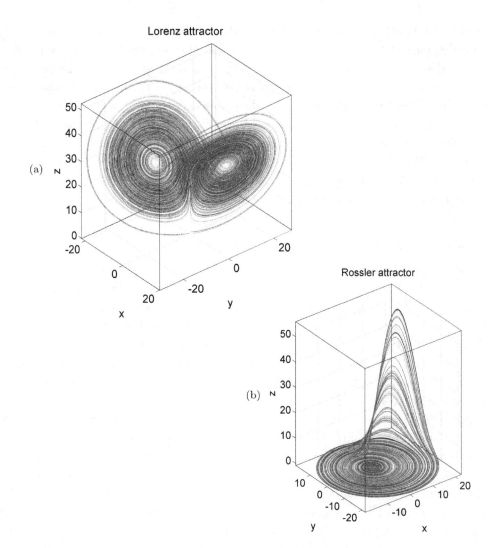

Fig. 5.6   The *Lorenz attractor* is a butterfly-shaped region in Cartesian x, y, z space that describes the asymptotic solution space of the Lorenz equations: $\frac{dx}{dt} = s(y-x)$, $\frac{dy}{dt} = x(r-z)-y$, $\frac{dz}{dt} = xy-bz$, with parameters: $s = 10, r = 28, b = \frac{8}{3}$. Regardless of the starting position, all trajectories gravitate to the attractor, although motion within it is chaotic. The Rössler *strange attractor* exhibits similar behaviour. The multi-coloured trajectories are the solution to the equation set: $\frac{dx}{dt} = -y - z$, $\frac{dy}{dt} = x + ay$, $\frac{dz}{dt} = z(x - c) + b$, with parameters: $a = \frac{1}{10}, b = \frac{1}{10}, c = 14$. For (a) and (b), *Runge–Kutta* numerical (fixed timestep) solvers are used to evaluate the equations.

of x, y, z is chaotic, nonetheless all trajectories tend to gravitate towards a particular butterfly-shaped surface. The 'Butterfly Effect' indeed, and an iconic image of *Chaos Theory*. The surface was later coined by Davide Ruelle (1935–) & Floris Takens (1940–2010) as a *Strange Attractor*.

Another chaotic system with a strange attractor was discovered by Otto Rössler (1940–), based upon the time variation of x, y, z variables defined by the following

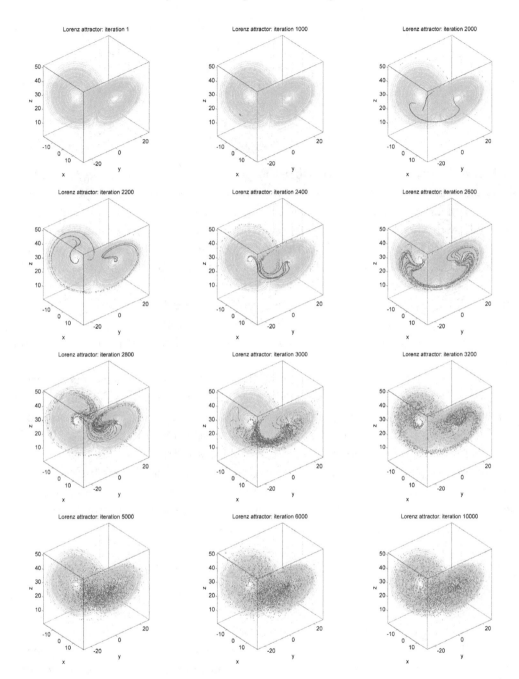

Fig. 5.7   Illustration of the properties of the Lorenz attractor, and a reworking by the author of Shaw's classic simulation [43]. Applying the Lorenz equations, a cluster of initial x, y, z values separated by a tiny random deviation will eventually spread out evenly throughout the *strange attractor*. In this simulation 5,000 distinct objects are initially separated by randomly distributed x, y, z values samples from range $[0.99, 1.01]$. The timestep for the Runge–Kutta solver is $\Delta t = 0.005$.

differential equations [45, p. 123]):

$$\frac{dx}{dt} = -y - z$$

$$\frac{dy}{dt} = x + ay$$

$$\frac{dz}{dt} = z(x - c) + b \tag{5.25}$$

Figure 5.6(b) was generated using the parameters: $a = \frac{1}{10}, b = \frac{1}{10}, c = 14$ and a *Runge–Kutta* solver was used to evaluate the 10,000 trajectory points with a timestep of $\Delta t = 0.01$. To demonstrate the strange attractor, 20 trajectories are overlaid, with random starting points with x, y, z values chosen from the interval $[-1, 1]$. The Rössler attractor is curious. Trajectories are mostly circular spirals in the x, y plane, but occasionally these rise up dramatically to form a sort of dorsal fin.

The Lorenz attractor in Fig. 5.6(a) was computed in a very similar fashion. Again only 20 different trajectories were chosen, mostly for aesthetic reasons, as too many makes the attractor hard to see when the trajectories, plotted using different colours, overlap. A timestep of $\Delta t = 0.005$ was used and $N = 5,000$ trajectory points were plotted.

Perhaps a more convincing demonstration of the Lorenz attractor is my reworking of a series of illustrations created by Shaw *et al.* in their *Chaos* paper in *Scientific American* in 1986. [43]. A plot of the Lorenz attractor is made and the trajectories of a cluster of initially very closely packed objects are computed and overlaid using the Lorenz equations. The progress of these objects is tracked via a series of video frames, and a selection of these is presented in Fig. 5.7. The objects initially stay in close proximity and move around the attractor together. However, as time progresses they eventually spread out and, after about 5,000 iterations, the objects cover the whole attractor, but *nowhere else* in the x, y, z space.

5.5 The Attraction of Attractors and Geometric Order from Chaos

One of the key ideas of *Chaos Theory* is to seek meaning from *diagrammatic* representations of all possible trajectories, in the form of a *Poincaré map*, rather than try to seek patterns in seemingly random motion with time. This is a refreshing perspective for analysis of dynamics, and can aid understanding of simpler, solvable systems, such as Simple Harmonic Motion of pendulums. The *strange attractors* of Lorenz and Rössler yield a weird and beautiful form of 'asymptotic order', that would otherwise be unseen if one just stuck to conventional time-series analysis. Underpinning this exploration is further use of numerical Calculus methods such as *Verlet* and *Runge–Kutta*, and also the mathematical magic of the *Euler–Lagrange* system of generating equations for multi-particle systems.

Chapter 6

A Standard Atmosphere

6.1 An Onion Skin Paradise

To quote Lloyd in *A Concise Guide to Weather* [26]:

> The atmosphere around our planet is critically important for our survival. It absorbs the energy we need, reflects the surplus back into space and provides a layer of protection against the harmful elements of incoming radiation. It contains gases, clouds, particles of dust and other particles called aerosols; the main gases being nitrogen and oxygen. Arranged in distinctive layers with no discernable outer limit, most of the atmospheric mass lies below an altitude of 100 km. Compared to the Earth's dimensions, the atmosphere is very thin: the equivalent of a layer of skin to an onion.

Within the first 11 km of this onion skin is our fragile paradise, the *Troposphere*. Nearly all of life is contained, and the vast majority of weather effects occur here. The major reason for this is the rapid decay of air pressure with altitude. Since gases are highly compressible, this means a greater density of air molecules are present at lower altitudes. According to Barry *et al.* [2, p. 32], the troposphere comprises about 75% of the total mass of the atmosphere, and nearly all the water vapour and aerosols. In this chapter we will develop what is known as the *International Standard Atmosphere* (ISA) model, with the goal of determining how average air pressure and temperature vary with altitude, and what effect humidity (i.e. the % saturation of water vapour) has on these variations.

6.2 The International Standard Atmosphere

The ISA is an idealised model of the variation of air pressure and temperature with altitude. The assumptions are as follows:

- The atmosphere comprises a number of fixed layers, each with a constant temperature gradient ('lapse rate') with altitude.

- The atmosphere consists of a single ideal 'air' gas, whose molecular mass takes into account the average compositions of different gases, e.g. nitrogen (78%), oxygen (21%), 0.9% argon, 0.04% carbon dioxide, etc.

At the base of the Troposphere, altitude $h = 0.0$ km, temperature $T_0 = (273 + 15)$ K, pressure $P_0 = 101,325$ Pa. The other layers of the ISA are defined in the following table[1]:

Level	Base h (km)	Lapse rate L (K/km)
Troposphere	0.0	−6.5
Tropopause	11.0	0.0
Stratosphere	20.0	1.0
Stratosphere	32.0	2.8
Stratopause	47.0	0.0
Mesosphere	51.0	−2.8
Mesosphere	71.0	−2.0

The ISA model is deemed valid until the *Mesopause* at about $h = 84.8$ km. An example ISA, assuming dry air (i.e. zero humidity) is illustrated in Fig. 6.1(a) for the first three atmospheric layers (Troposphere, Tropopause, Stratosphere). In the temperature vs altitude graph the constant lapse rates correspond to the fixed gradients. Pressure smoothly decays from 101,325 Pa to about 1 Pa at the top of the Stratosphere. The predicted variation depends on the lapse rate, and the equations for pressure vs altitude will be derived in the next few sections. Figure 6.1(b) extends the model to the entire ISA. Note although the average temperature in the Stratosphere rises due to a positive lapse rate, you would not derive much warmth at the start of an extreme altitude free-fall like Eustace in 2014, or Baumgartner in 2012. This is because the air density in the Stratosphere, and therefore the rate of exchange of air molecule kinetic energy with your body, would be very small compared to similar temperatures in the Troposphere.

6.3 Variation of Atmospheric Pressure vs Altitude

Consider a $1 \, \text{m}^2$ horizontal cross-section parcel of air of density r at an altitude z, with vertical width dz. As illustrated in Fig. 6.2(a), the atmospheric pressure change

---

[1]The table is broadly in agreement with the ISA described in https://en.wikipedia.org/wiki/ International_Standard_Atmosphere (accessed 10/4/2020). Note there are other more extensive models (i.e. which extend to higher altitudes). The International Civil Aviation Organization (ICAO) maintain a reference up to 80 km, and airspeed indicators are calibrated to this standard. NRLMSISE-00 and JB2008 are published standards by US Navy and Space-Force which are used to predict atmospheric drag on orbiting satellites. They have an altitude range of up to 1,000 km.

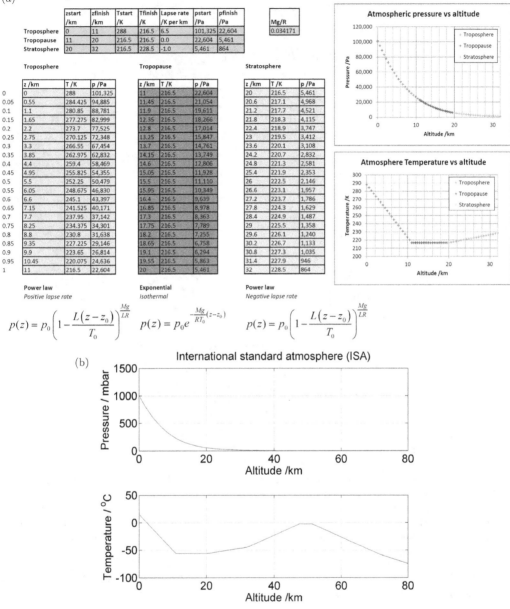

$$p(z) = p_0 \left(1 - \frac{L(z - z_0)}{T_0}\right)^{\frac{Mg}{LR}} \qquad p(z) = p_0 e^{-\frac{Mg}{RT_0}(z - z_0)} \qquad p(z) = p_0 \left(1 - \frac{L(z - z_0)}{T_0}\right)^{\frac{Mg}{LR}}$$

Fig. 6.1   An example ISA, assuming dry air (i.e. zero humidity). (a) illustrates a spreadsheet calculation for the first three atmospheric layers, the Troposphere, the Tropopause and the Stratosphere. In the temperature vs altitude graph the constant lapse rates correspond to the fixed gradients of the T vs altitude line segments. Atmospheric pressure P smoothly decays from 101,325 Pa to about 1 Pa at the top of the Stratosphere. The predicted variation depends on the lapse rate, and the equations for pressure vs altitude are either power-laws (non-zero but fixed lapse rate) or exponential decays based upon a Boltzmann-factor (zero lapse rate, i.e. isothermal atmosphere as in the Tropopause). (b) extends the model to the entire ISA.

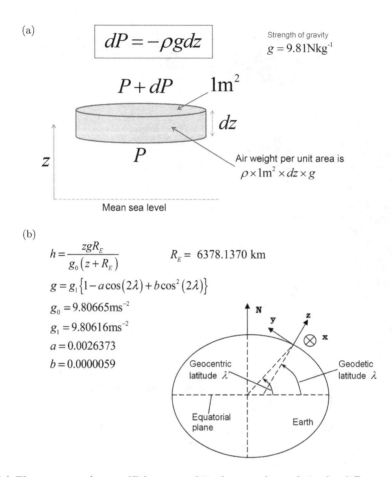

Fig. 6.2 (a) The pressure change dP between altitudes z and $z + dz$ is the difference in weight of air per unit area. i.e. $dP = -\rho g dz$ where ρ is the air density and g is the strength of gravity. The minus sign is required such that pressure reduces with altitude. Why? Well the air column *from altitude z upwards* increases in weight as z decreases. So as z reduces, the weight of air above per unit area (i.e. the air pressure) increases. (b) To account for changes in g due to the ellipsoidal shape of the Earth (e.g. the standard WGS-84), let us define a *geopotential altitude* h in terms of actual altitude z, measured perpendicular to a tangent plane to the WGS-84 ellipsoid.

dP between altitudes z and $z + dz$ resulting from the removal of the air parcel from the total weight of air above is:

$$dP = -\rho g dz \tag{6.1}$$

Let us assume that the air column is comprised of dry air with molar mass $M = 28.96\,\text{gmol}^{-1}$. Assume that air is an *ideal gas*.[2] If n molecules occupies volume

---

[2] An *ideal gas* assumes high-speed random motion of molecules, which have a *mean free path* between collisions that is much greater than molecular dimensions. In an ideal gas we can ignore the effects of molecular size (an ideal gas is infinitely compressible), and we can also ignore attraction or

V at pressure P at (absolute) temperature T then:

$$PV = nRT \tag{6.2}$$

where R is the molar gas constant $R = 8.414\,\mathrm{Jmol^{-1}K^{-1}}$.

The density of the air is the mass of n moles divided by the volume V:

$$\rho = \frac{nM}{V} \tag{6.3}$$

Hence

$$\rho = \frac{nM}{\frac{nRT}{P}} = \frac{MP}{RT} \tag{6.4}$$

Substituting $\rho = \frac{MP}{RT}$ into $dP = -\rho g\,dz$ yields an integral expression for the pressure with altitude z. i.e. since $dP = -\frac{MP}{RT}g\,dz$:

$$\int_{P_0}^{P} \frac{1}{P'}dP' = -\frac{M}{R}\int_{z_0}^{z} \frac{g(z')}{T(z')}dz' \tag{6.5}$$

Note the Earth is not perfectly spherical and is more precisely modelled as an *ellipsoid*.[3] To account for changes in gravitational field strength g with latitude, let us firstly define a *geopotential altitude* h such that:

$$\therefore \int_{P_0}^{P} \frac{1}{P'}dP' = -\frac{Mg}{R}\int_{h_0}^{h} \frac{1}{T(h')}dh'$$

$$\therefore \ln\left(\frac{P}{P_0}\right) = -\frac{Mg}{R}\int_{h_0}^{h} \frac{1}{T(h')}dh' \tag{6.6}$$

where g is a function of *geodetic* latitude λ (see Fig. 6.2)

$$g = g_1\left(1 - a\cos 2\lambda + b\cos^2 2\lambda\right)$$

$$g_0 = 9.81665\,\mathrm{Nkg^{-1}}; \quad g_1 = 9.80616\,\mathrm{Nkg^{-1}} \tag{6.7}$$

$$a = 0.0026373; \quad b = 0.0000059$$

Actual altitude z (i.e. the vertical extent, normal to the ellipsoidal surface of the Earth) is related to h, g and the Earth radius by:

$$z = \frac{R_\oplus h g_0}{g R_\oplus - g_0 h} \tag{6.8}$$

---

repulsion due to the distribution of charge in molecules. All collisions are *elastic*, and the energy of the system is deemed to be 100% kinetic.

[3]The Earth is more precisely modelled as an *ellipsoid* than a perfect sphere, although this is a small modification compared to Earth dimensions. The difference between the polar radius (6,356.75 km) and the equatorial radius (6,378.14 km) is just 21.39 km. Note this is comparable to the altitude gain (19.78 km) from *Challenger Deep* in the Marianas Trench (the lowest point on the surface of the Earth), to the summit of Mount Everest, the highest point on the surface of the Earth, as measured from mean sea level.

The mean equatorial radius of the Earth is $R_\oplus = 6378.1370\,\text{km}$.

If one is happy to assume the approximation of a spherical Earth of radius $R_\oplus$, and ignore the reduction of the strength of gravity with altitude, then $g = 9.81\,\text{Nkg}^{-1}$ and $h = z$.

6.3.1 *Special case #1: Isothermal atmosphere $T = T_0$, zero humidity*

This special case is valid in the *Tropopause* and the *Stratopause*, i.e. when the lapse rate is zero. Factoring out the (constant) temperature T_0, Eq. (6.6) becomes:

$$\ln\left(\frac{P}{P_0}\right) = -\frac{Mg}{RT_0}\int_{h_0}^{h} dh'$$

$$\ln\left(\frac{P}{P_0}\right) = -\frac{Mg(h - h_0)}{RT_0} \tag{6.9}$$

$$P(h) = P_0 e^{-\frac{Mg(h-h_0)}{RT_0}} \tag{6.10}$$

i.e. for an isothermal atmosphere, the pressure varies with altitude h via a *Boltzmann factor* $e^{-\frac{\text{energy}}{\text{thermal energy}}}$ relationship, since the energy in this case is the change in gravitational potential energy (GPE) $\Delta E = Mg(h - h_0)$. A plot of the exponential decay of pressure with altitude is illustrated in Fig. 6.3(b).

6.3.2 *Special case #2: Constant, but non-zero, lapse rate L, zero humidity*

As in the ISA model of the Troposphere and Stratosphere, assume temperature T varies linearly with altitude h, i.e. with a constant lapse rate L.

$$T(h) = T_0 - L(h - h_0) \tag{6.11}$$

Equation (6.6) now becomes:

$$\ln\left(\frac{P}{P_0}\right) = -\frac{Mg}{R}\int_{h_0}^{h}\frac{1}{T(h')}dh'$$

$$\ln\left(\frac{P}{P_0}\right) = -\frac{Mg}{R}\int_{h_0}^{h}\frac{1}{T_0 - L(h' - h_0)}dh'$$

$$\ln\left(\frac{P}{P_0}\right) = \frac{Mg}{LR}\int_{h_0}^{h}\frac{-L}{T_0 - L(h' - h_0)}dh'$$

$$\ln\left(\frac{P}{P_0}\right) = \frac{Mg}{LR}\Big[\,\ln\left(T_0 - L(h' - h_0)\right)\,\Big]_{h_0}^{h}$$

$$\ln\left(\frac{P}{P_0}\right) = \frac{Mg}{LR}\ln\left(\frac{T_0 - L(h - h_0)}{T_0}\right)$$

$$\ln\left(\frac{P}{P_0}\right) = \ln\left(\left(\frac{T_0 - L(h - h_0)}{T_0}\right)^{\frac{Mg}{LR}}\right) \tag{6.12}$$

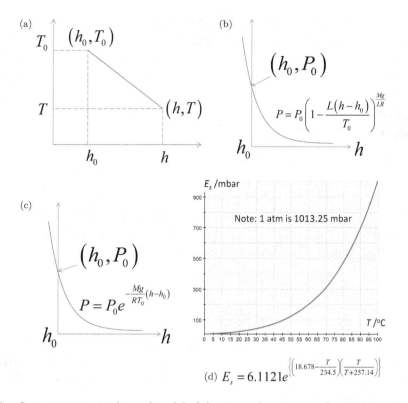

Fig. 6.3 A refinement to an *isothermal* model of the atmosphere is one where temperature T varies in a *linear* fashion with increasing altitude h, as shown in (a). In the *Troposphere*, temperature diminishes by a *lapse rate* of between 0.5 and 1.0°C (or Kelvin) for every 100 m of altitude gain. For dry air, i.e. the effect of humidity can be ignored, this means pressure P will decay in a *power law* fashion with altitude, as depicted in (b). A special case, illustrated in (c), is when the *lapse rate is zero*, which results in an *exponential* decay. E_s, the saturation vapour pressure of atmospheric gases, is plotted vs temperature in (d), using the empirical *Arden-Buck* equation. The contribution to atmospheric pressure due to water vapour is $P_u = UE_s$ where U is the humidity. $U = 0$ for dry air whereas $U = 1$ for saturated air, i.e. in conditions of '100% humidity'.

The variation of pressure P with altitude h is therefore a *power law*, with the power being the dimensionless quantity $\frac{Mg}{LR}$.

$$P(h) = P_0 \left(1 - \frac{L(h - h_0)}{T_0}\right)^{\frac{Mg}{LR}} \tag{6.13}$$

Note if we express lapse rate $L = \frac{\Delta T}{\Delta h}$, then $\frac{Mg}{LR} = \frac{Mg\Delta h}{R\Delta T}$ which is a kind of 'fractional Boltzmann factor'.

6.4 The Effect of Humidity

6.4.1 *Variation of atmospheric density with temperature, pressure and humidity*

Any sensible description of non-arid *climatology* (i.e. the study of the atmosphere) will need to take into account the impact of variable amounts of water vapour contained within the air column. In the troposphere at least, the presence of water vapour can have a significant influence upon thermodynamic variables such as temperature and pressure, and is obviously a fundamental component of weather phenomena such as cloud and fog. To model the effect of humidity upon temperature and pressure, let us modify our original single ideal gas assumption to consider a composite of dry air and water vapour. The molar masses of dry air and water vapour are, respectively:

$$M_d = 28.96 \, \text{gmol}^{-1} \tag{6.14}$$

$$M_v = 18.02 \, \text{gmol}^{-1} \tag{6.15}$$

The air pressure P is defined to be the sum of the pressures associated with dry air (d) and water vapour (v) components. If these are ideal gases of temperature T, which comprise n_d and n_v moles each per volume V:

$$P = P_d + P_v \tag{6.16}$$

$$P = \frac{n_v RT}{V} + \frac{n_d RT}{V} \tag{6.17}$$

Let U be the proportion of the *saturation vapour pressure* E_s, which is the maximum vapour pressure before the water liquifies (and therefore ceases to be an atmospheric gas). The range of U is $0 \leq U \leq 1$, with $U = 1$ corresponding to 100% humidity. Hence water vapour pressure P_v is:

$$P_v = U E_s \tag{6.18}$$

An empirical relationship between E_s and temperature (in degrees Celsius not K, i.e. hence the c subscript) is the *Arden–Buck equation*,[4] which is plotted in Fig. 6.3(d). Note (Kelvin) temperature is $T = T_c + 273$. Note the empirical formula for E_s is calculated in mbar. 1 mbar = 100 Pa, so multiply E_s by 100 if P is in Pa.

$$E_s = 6.1121 e^{\left(18.678 - \frac{T_c}{234.5}\right)\left(\frac{T_c}{T_c + 257.14}\right)} \tag{6.19}$$

The overall humid air density can be computed from a sum of contributions from dry air of molar mass M_d and water vapour of molar mass M_v:

$$\rho = \frac{n_v M_v + n_d M_d}{V} = \frac{M_d}{V}\left(n_d + n_v \frac{M_v}{M_d}\right) \tag{6.20}$$

From the ideal gas equations:

$$n_v = \frac{V}{RT}P_v = \frac{V}{RT}U E_s$$

$$n_d = \frac{V}{RT}P_d = \frac{V}{RT}(P - P_v) = \frac{V}{RT}(P - U E_s) \tag{6.21}$$

---

[4]https://en.wikipedia.org/wiki/Arden_Buck_equation.

Therefore,

$$\rho = \frac{M_d}{V}\left(n_d + n_v \frac{M_v}{M_d}\right)$$

$$\rho = \frac{M_d}{V}\left(\frac{V}{RT}(P - UE_s) + \frac{V}{RT}UE_s\frac{M_v}{M_d}\right)$$

which results in an expression for air density ρ involving overall pressure P, saturation vapour pressure E_s, absolute temperature T, relative humidity U, and the dry air and water vapour molecular masses M_d and M_v.

$$\rho = \frac{M_d}{RT}\left(P - U\left(1 - \frac{M_v}{M_d}\right)E_s\right) \tag{6.22}$$

6.4.2 *Effect of humidity upon lapse rate*

The effect of humidity upon lapse rate $L = -\frac{dT}{dh}$ is significant, and can be modelled by the following equation:[5]

$$L = g\frac{1 + \frac{rL_{vap}}{R_{sd}T}}{c_{pd} + \frac{(L_{vap})^2 r}{R_{sw}T^2}} \tag{6.23}$$

$$r = \frac{R_{sd}}{R_{sw}}\frac{UE_s}{P - UE_s} \tag{6.24}$$

where the *specific latent heat of vaporization* of water $L_{vap} = 2,501\,\mathrm{kJkg^{-1}}$, $R_{sd} = \frac{R}{M_d}$ is the *specific gas constant for dry air*: $R_{sd} = \frac{8.314\,\mathrm{Jmol^{-1}K^{-1}}}{0.02896\,\mathrm{kgmol^{-1}}} \approx 287\,\mathrm{Jkg^{-1}K^{-1}}$. $R_{sw} = \frac{R}{M_v}$ is the *specific gas constant for water vapour*: $R_{sw} = \frac{8.314\,\mathrm{J\,mol^{-1}K^{-1}}}{0.01802\,\mathrm{kgmol^{-1}}} \approx 461.5\,\mathrm{Jkg^{-1}K^{-1}} \cdot c_{pd} = 1003.5\,\mathrm{Jkg^{-1}K^{-1}}$ is the *specific heat capacity* (at constant pressure) of dry air, and g is the gravitational field strength determined by Eq. (6.7). In this equation, temperature T is in kelvin, and P is atmospheric pressure. U is the humidity fraction $0 \leq U \leq 1$. Note $r = \frac{R_{sd}}{R_{sw}}\frac{UE_s}{P-UE_s}$ is essentially the mass fraction of water vapour to dry air.

6.4.3 *Variation of dew point with temperature*

The *dew point* is the temperature T_d (in degrees celsius) to which a given parcel of air must be cooled, at constant barometric pressure, for water vapour to condense into water. The *August–Roche Magnus approximation*[6] defines the dew point in terms of relative humidity U (with values $0\ldots1$) and ambient air temperature T_c (in degrees

---

[5]https://en.wikipedia.org/wiki/Lapse_rate (accessed 11/4/2022).
[6]http://andrew.rsmas.miami.edu/bmcnoldy/Humidity.html (accessed 10/4/2020).

Celsius).

$$T_d = \frac{b\left(\ln U + \frac{aT_c}{b+T_c}\right)}{a - \ln U - \frac{aT_c}{b+T_c}} \tag{6.25}$$

$$a = 17.625; \quad b = 243.04 \tag{6.26}$$

6.4.4 *Variation of water boiling point with atmospheric pressure*

To boil water, it must undergo a *phase transition* from liquid to gas. This requires a certain amount of heat, the latent heat of vaporization, to break the inter-molecular bonds inherent in the water. The gradient of a phase transition line in a P, T diagram is given by the *Clausius–Clapeyron equation*:

$$\frac{dP}{dT} = \frac{M_v L_{\text{vap}}}{T \Delta V} \tag{6.27}$$

where M_v is the molar mass of water and $L_{\text{vap}} = 2,501\,\text{kJkg}^{-1}$ is the specific latent heat of vaporization, i.e. the energy required to change state from liquid to gas, assuming the system is at boiling temperature. ΔV is the volume change (per mole) between liquid and gaseous states. In a liquid to gas transition we can assume the volume change is sufficiently large as to ignore the original fluid volume. If one assumes the resulting gas is ideal, then the volume change per mole is $\Delta V = \frac{RT}{P}$. The *Clausius–Clapeyron* equation then becomes:

$$\frac{dP}{dT} = \frac{M_v L_{\text{vap}}}{RT^2} P \tag{6.28}$$

$$\int_{P_*}^{P} \frac{1}{P'} dP' = \frac{M_v L_{\text{vap}}}{R} \int_{T_*}^{T} \frac{1}{T'^2} dT' \tag{6.29}$$

$$\ln\left(\frac{P}{P_*}\right) = -\frac{M_v L_{\text{vap}}}{R}\left(\frac{1}{T} - \frac{1}{T_*}\right) \tag{6.30}$$

This yields an expression for the boiling point of water T_{boil} with atmospheric pressure P, given a known pairing (T_*, P_*), such as $P_* = 101,325\,\text{Pa}$, $T_* = 373\,\text{K}$. i.e. one *atmosphere* of atmospheric pressure, and a water temperature of 100°C.

$$\therefore T_{\text{boil}} = \left(\frac{1}{T_*} - \frac{R}{M_v L_{\text{vap}}}\ln\left(\frac{P}{P_*}\right)\right)^{-1} \tag{6.31}$$

6.4.5 *A humid atmosphere model of the troposphere*

The modifications due to humidity described in the previous sections can be used to construct a more precise model of the variation of pressure P, temperature T, lapse rate L (and also dew point T_d and water boiling point T_{boil}) with altitude z. Unlike the dry-atmosphere special cases described above, an analytic solution is not possible. Instead a numerical (Euler) scheme is used.

Initial conditions:

$$P_0 = P_* = 1,013.25 \,\text{mbar}, \; T_* = 373 \,\text{K}, \; T_0 = 15°\text{C} = 288 \,\text{K} \tag{6.32}$$

Note: $1 \,\text{mbar} = 100 \,\text{Pa}$.

Inputs: Altitude z (km), humidity U where $0 \leq U \leq 1$, geodetic latitude λ (which is 51.5°N for London, UK).

Numerical recipe:

- Determine geopotential height h, from a rearrangement of Eqs. 6.7 and 6.8

$$h = \frac{g}{g_0} \frac{R_\oplus z}{R_\oplus + z} \tag{6.33}$$

$$g = g_1 \left(1 - a \cos 2\lambda + b \cos^2 2\lambda\right)$$

$$g_0 = 9.81665 \,\text{Nkg}^{-1}; \quad g_1 = 9.80616 \,\text{Nkg}^{-1} \tag{6.34}$$

$$a = 0.0026373; \quad b = 0.0000059; \quad R_\oplus = 6378.1370 \,\text{km}$$

- Define a linearly spaced set of altitudes $0 \leq x \leq h$ in steps of $\Delta x = 0.01$ km

$$x_{n+1} = x_n + \Delta x \tag{6.35}$$

- Set pressure and temperature to be P_0, T_0 respectfully, at $x_0 = 0$ km.

Iterative loop, i.e. while $x_n \leq h$:

- Calculate saturation vapour pressure E_s, lapse rate L, air density ρ, water boiling point T_{boil} and dew point T_d. Note Celsius temperature $T_c = T - 273$. $L_{\text{vap}} = 2,501 \,\text{kJkg}^{-1}$, $R_{sd} = \frac{R}{M_d}$ is the *specific gas constant for dry air*: $R_{sd} = \frac{8.314 \,\text{Jmol}^{-1}\text{K}^{-1}}{0.02896 \,\text{kgmol}^{-1}} \approx 287 \,\text{Jkg}^{-1}\text{K}^{-1}$. $R_{sw} = \frac{R}{M_v}$ is the *specific gas constant for water vapour*: $R_{sw} = \frac{8.314 \,\text{Jmol}^{-1}\text{K}^{-1}}{0.01802 \,\text{kg mol}^{-1}} \approx 461.5 \,\text{Jkg}^{-1}\text{K}^{-1}$. $c_{pd} = 1003.5 \,\text{Jkg}^{-1}\text{K}^{-1}$ is the *specific heat capacity* (at constant pressure) of dry air.

$$\frac{E_s}{\text{mbar}} = 6.1121 e^{\left(18.678 - \frac{T_c}{234.5}\right)\left(\frac{T_c}{T_c + 257.14}\right)} \tag{6.36}$$

$$L = g \frac{1 + \frac{r L_{vap}}{R_{sd} T}}{c_{pd} + \frac{(L_{vap})^2 r}{R_{sw} T^2}} \tag{6.37}$$

$$r = \frac{R_{sd}}{R_{sw}} \frac{U E_s}{P - U E_s} \tag{6.38}$$

$$\rho = \frac{M_d}{RT} \left(P - U \left(1 - \frac{M_v}{M_d}\right) E_s\right) \tag{6.39}$$

$$T_d = \frac{b \left(\ln U + \frac{a T_c}{b + T_c}\right)}{a - \ln U - \frac{a T_c}{b + T_c}}; \quad a = 17.625; \quad b = 243.04 \tag{6.40}$$

$$T_{boil} = \left(\frac{1}{T_*} - \frac{R}{M_v L_{vap}} \ln\left(\frac{P}{P_*}\right)\right)^{-1} \tag{6.41}$$

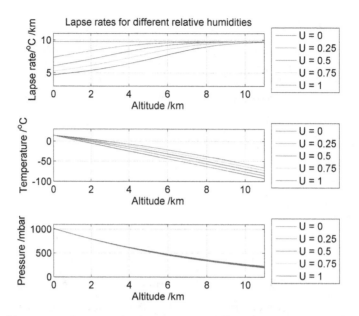

Fig. 6.4   Plot of lapse rate L, atmospheric temperature T_c and pressure P vs altitude, computed by a humid atmospheric model coded using MATLAB. Initial conditions are: $P_0 = 1{,}013.25$ mbar, $T_0 = 15°C$, and the model employs an Euler numerical scheme, cumulatively applying discrete pressure and temperature changes every 0.01 km altitude gain.

Note: If pressures P and E_s are in mbar, you'll need to multiply by 100 to yield a density ρ in kgm^{-3}. Also, T_d is calculated in degrees Celsius, whereas all other instances of temperature T are in Kelvin.

• Now calculate the change in pressure ΔP and change in temperature ΔT

$$\Delta P = -\rho g \Delta z \approx -\rho g \Delta x \tag{6.42}$$

$$\Delta T = -L \Delta x \tag{6.43}$$

• Update temperatures and pressures and then iterate until $x = h$

$$T_{n+1} = T_n + \Delta T \tag{6.44}$$

$$P_{n+1} = P_n + \Delta P \tag{6.45}$$

• Once $x_n = h$ (or the next altitude step would exceed h), output P, T, E_s, $L, \rho, T_d, T_{\text{boil}}$.

The iteration above can be encoded into a MATLAB (or equivalent computer programming language function) and therefore a plot of $P, T, E_s, L, \rho, T_d, T_{\text{boil}}$ vs altitude z and humidity U can be constructed, as illustrated in Figs. 6.4 and 6.5.

For example, a MATLAB function like:

```
[P(n,m),T(n,m),L(n,m),Tdew(n,m),Tboil(n,m)] =
PT_model(z(n),U(m),P0,T0,Pboil0,Tboil0,lat);
```

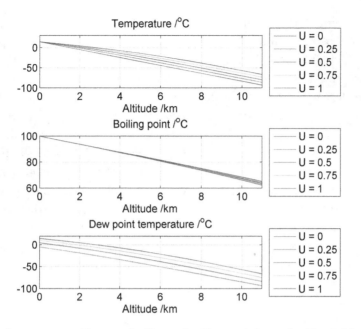

Fig. 6.5 Plot of temperature T_c, water boiling point T_{boil} and dew point T_d vs altitude, computed by a humid atmospheric model coded using MATLAB. Initial conditions are: $P_0 = 1{,}013.25$ mbar, $T_0 = 15°C$, and the model employs an Euler numerical scheme, cumulatively applying discrete pressure and temperature changes every 0.01 km altitude gain.

This would calculate the matrix element n, m of arrays of output parameters, based upon input vectors of altitude z and humidity U. It is the columns of these matrices that will then be plotted. Different colours correspond to different humidities.

In Fig. 6.4, we can see that humidity has quite a profound effect upon lapse rate, and the mountaineering rule-of-thumb of '1°C per 100 m height gain in the dry, 0.5°C per 100 m heigh gain in the wet' seems justified. Although at low altitudes at 100% humidity, the lapse rate can drop to under 0.5°C per 100 m. However, the effect of humidity on air pressure is less significant. This is because the contribution of vapour pressure at modest temperatures (i.e. below T_0) to the total air pressure is small, as modelled by the Arden–Buck equation for E_s, which is plotted in Fig. 6.3(d). As a consequence, the trends of Fig. 6.5 are not surprising. The effect of humidity on water boiling point is small, whereas the effect of humidity on dew point is much more significant.

6.5 Raise the Standard, and Embrace Empiricism

A Standard Atmosphere is intended as another, hopefully readily tangible, context to incentivise the acquisition of mathematical modelling skills. As in previous sections, there is an emphasis on the use of numerical methods, and in this case we used an iterative approach to compute the air pressure, temperature, density, lapse rate, boiling point and dew point at different altitudes. There is also scope for analytical work if one ignores the effect of humidity, i.e. consider a perfectly arid atmosphere. A difference between this chapter and the previous is the thermodynamic context, and also the use of *empirical* relations. Although it is nearly always preferable to derive algebraic relationships between physical quantities from 'first principles', there are certain situations (for example the temperature variation of saturation vapour pressure E_s with temperature) which are so complicated, that to make progress we must go 'in ignorance' to measured data and distil any correlations directly, perhaps with only an educated guess of the mathematical form of the relationship between variables. In a practical scenario like modelling the properties of the atmosphere, it is more important for aeronautic engineering applications to *get a realistic answer*. If we restrict ourselves to problems that we can solve analytically, such as a $U = 0$ perfectly arid air column, physics is less helpful in the real world.

Chapter 7

The Subtlety of Rainbows

7.1 Arts and Science, Not Versus

In 1820, the English poet John Keats published the poem *Lamia*, with the inference that a scientific deconstruction of the sublime beauty of a rainbow would somehow 'empty the haunted air'. As beautiful as the poem is, I think the anti-science stance requires a pretty strong riposte. I think we could afford the evolution of complex life, and other exquisite natural structures during the 13.8 billion years of a possibly infinite Universe, a bit more scope for scholarly inspection and source of wonder! C. P. Snow's 1959 *The Two Cultures* Rede Lecture describes the inevitable communications problems that arise from a scientifically snooty, humanities-dominated elite running a post-war, *technology dominated* society. Have we heeded Snow's warnings in the past 60 years? We shall let history judge the Latin-quipping British government's response to the 2020 coronavirus pandemic, a problem both identified and solvable by the rigorous appliance of Science. Whether there is a cultural phase separation or not, I think there is potential for a positive spin on the Two Cultures hypothesis. As a professional scientist, *and* a lover of art and literature, I would like to echo Richard Feynman's excellent counter to the sentiment of *Lamia* that he describes on page 2 of *The Pleasure of Finding Things Out* [9]. In essence, Feynman contrasts his view of a flower compared to one of his purely artistic friends, who thinks his desire to deconstruct 'dulls' the essence of the flower. In Feynman's view, knowledge about the processes and microstructure associated with the flower only *adds* to its beauty, and connects it with the ecosystem which produces it.

As a whimsical counterpoint to Keats, and those who still pursue the nonsense of an arts vs science narrative (it is *and* not vs!) I present a delightful mathematical model of the optical effect we know from childhood as a *rainbow*. The model is based upon the ideas of the French philosopher Réne Descartes (1596–1650), and the mathematical narrative in Rees' excellent book *Physics by Example* [38]. As with much of Physics, the model relates what we can readily observe on a macroscopic scale (the rainbow) to a hidden microscopic world (the raindrop), using universal principles (wave nature of light, reflection and refraction) which in turn connect the rainbow phenomenon to a much broader scope of optical systems. You just can't do this simply by labelling phenomena. A mathematical model also enables you to make precise predictions regarding the elevation angle of a rainbow (about 41.7° for a primary bow, with an angular width of about 1.6°), and indeed provides a recipe how you could make one artificially using a mist generator and a light source behind you.

The Romantics sought meaning in the natural world based upon how it made them feel, and the stirring of the emotions when encountering the sublime. There is nothing wrong with this in my view (I spend much of my holidays climbing mountains for precisely this reason), but it is obviously *introspection* rather than a more universal interrogation of reality; of what lies beyond the direct experience of poetic *homo sapiens*. There is of course a delightful irony inherent in even the most dispassionate scientific description of a rainbow. As we shall soon demonstrate, the observer forms the conclusion of the optical system, and cannot be separated from it. The perception of the coloured bands is a consequence of the frequency content of the light at the retina of the observer, resulting from the light path from the Sun to a raindrop to the eye. In other words, there is no objective reality of a rainbow independent of the observer, and certainly no pot of gold beyond it. In fact it forms circles if uninterrupted by topography. The rainbow is, quite literally, *in* the eye of the beholder.

7.2 Descartes' Theory of the Rainbow

7.2.1 *Assumptions*

Our 'Cartesian' model of a rainbow is based upon the following ideas:

- The source of light is from the Sun. This is such a long way away from the Earth that light travels in parallel rays. To produce a rainbow (as opposed to other similar optical effects), the light source must be in the *anti-solar* direction, i.e. behind the observer, as shown in Fig. 7.1.
- Light from the Sun is internally reflected off the interior of spherical raindrops, and the emerging light can be ray-traced to the eye of the observer. A single *internal reflection* creates the standard rainbow effect, but multiple reflections are also possible and these result in double, and higher-order rainbows.[1]
- All raindrops are perfectly spherical. Since water molecules are polar, electric fields can distort the spherical shape of rain droplets so this assumption may become invalid during thunderstorms.

---

[1] According to https://earthsky.org/earth/first-ever-image-of-5th-order-rainbow (accessed Jan 2019) *fifth order* rainbows (i.e. five internal reflections inside a raindrop) have been observed, but these are hard to see with a mark-one eyeball.

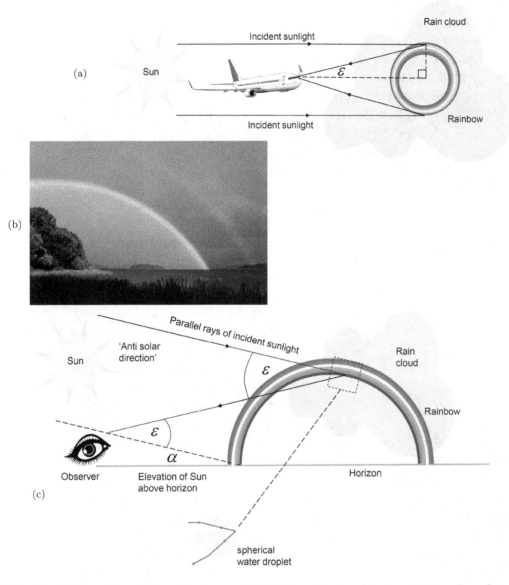

Fig. 7.1 (a) A rainbow is formed when sunlight in the anti-solar direction is internally reflected within a water droplet. The refractive index of water varies slightly with the frequency of light, and therefore some dispersion of colour is observed. Certain colours are preferentially concentrated at particular elevations angles ε to the observer. (b) *Alexander's Dark Band*, which is a darkening of the sky between the primary (inner) and secondary (outer) bows. Notice that the colour order between the primary and secondary bows is reversed. (c) If the Sun is at a high elevation angle α, topography may obscure part of the rainbow. When uninterrupted, as in the aircraft view (a), a rainbow should be a complete circle. (b) Photo Credit: Alexis Dworsky. CC by 2.0 DE. Wikipedia: Rainbow.

- Assume raindrops have identical liquid content (water) whose refractive index variation with optical frequency follows a defined semi-empirical model. This implies the raindrops are fairly uniform in temperature, pressure and impurity content. We will be using a model of pure water to compute the refractive index, so we shall neglect any effect of impurities.
- The refractive index of the ambient air is unity, i.e. the difference between light speed in air and light speed in a vacuum ($c = 2.998 \times 10^8$ ms^{-1}) can be neglected.
- Assume a *ray* model of sunlight propagation through the raincloud, i.e. don't include any interference effects resulting from the wave nature of light. For this assumption to be valid the wavelength of light within the raindrop $\lambda \ll a$, where a is the radius of the raindrops. Note wavelength of light (in air) varies between 400 nm (violet) and 750 nm (red).

7.2.2 The primary rainbow

A rainbow is formed by the deflection and focus of incident sunlight by spherical raindrops into a narrow range of elevation ε. The deflection occurs via internal reflection of incident sunlight, and the focusing effect results from the observation that the total angle of deflection passes through an *extremum* as the angle θ between light ray incidence and raindrop surface normal is varied. An *extremum* means a graph of ε vs θ has either a maximum or a minimum at a particular angle θ. The large multitude of raindrops within a rain cloud allows us to assume that all possible angles of θ are explored. If a mathematical model can determine $\varepsilon(\theta)$ then the postulate of 'extremum deflection' implies the first derivative of ε must pass through zero. When $\frac{d\varepsilon}{d\theta}$ is small, this means ε varies little as θ changes. Hence a wide range of sunlight rays (each with a different θ value) will result in similar ε angles. This is another way of saying sunlight is focused into a narrow range of elevation angles. Since the refractive index of water varies with frequency (see Fig. 7.4) we would expect each colour to form an arc at a slightly different elevation.

We can compute an expression for $\varepsilon(\theta)$ using the geometrical construction in Fig. 7.2. The construction was created noting all surface normals on a sphere point radially outwards, and thus emanate from the centre of the sphere. This means the reflected ray paths bound a pair of joined isosceles triangles. Also assumed is the Law of Reflection, that is the angle of incidence of light upon a surface equals the angle of reflection from that surface. (Where both angles are measured from the ray directions to the surface normal.) If we decompose the total deflection angle $\pi - \varepsilon$ (in radians) into (A) surface refraction at incidence + (B) internal reflection + (C) surface refraction at ray emergence:

$$\pi - \varepsilon = \underbrace{\theta - \phi}_{A} + \underbrace{\pi - 2\phi}_{B} + \underbrace{\theta - \phi}_{C} \tag{7.1}$$

$$\therefore \varepsilon = 4\phi - 2\theta \tag{7.2}$$

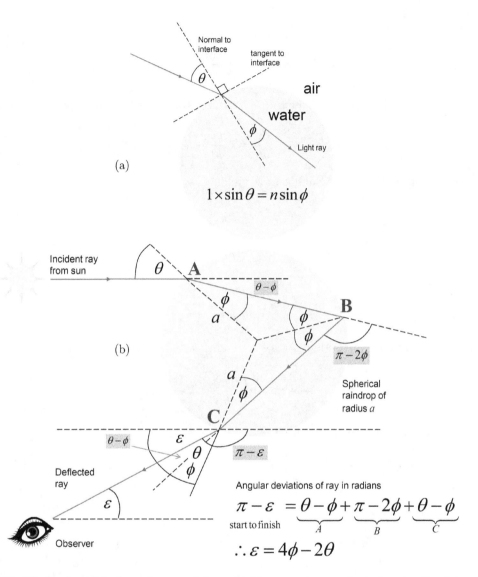

(a)

$$1 \times \sin\theta = n\sin\phi$$

(b)

Fig. 7.2 Idealised model of a primary rainbow. Light rays enter a spherical water droplet and are refracted by angle ϕ to the surface normal. These are internally reflected and then exit at an elevation angle ε. By tracing the ray path and use of Snell's law one can show $\varepsilon = 4\sin^{-1}\left(\frac{\sin\theta}{n}\right) - 2\theta$ where n is the refractive index. $\varepsilon(\theta)$ has a *maxima*, which explains the formation of a rainbow at $\varepsilon = 4\sin^{-1}\left(\sqrt{\frac{4-n^2}{3}} \times \frac{1}{n}\right) - 2\sin^{-1}\sqrt{\frac{4-n^2}{3}}$ corresponding to the light colour associated with n.

Snell's Law of refraction relates angle ϕ to refractive index n (recall the refractive index in air is deemed to be unity)

$$n\sin\phi = 1 \times \sin\theta \tag{7.3}$$

$$\therefore \sin\phi = \frac{\sin\theta}{n} \tag{7.4}$$

Hence:

$$\varepsilon = 4\sin^{-1}\left(\frac{\sin\theta}{n}\right) - 2\theta \tag{7.5}$$

The derivative of ε with respect to θ is perhaps most easily found from $\varepsilon = 4\phi - 2\theta$:

$$\frac{d\varepsilon}{d\theta} = 4\frac{d\phi}{d\theta} - 2 \tag{7.6}$$

Implicit differentiation of Snell's Law yields:

$$\sin\phi = \frac{\sin\theta}{n} \tag{7.7}$$

$$\cos\phi\frac{d\phi}{d\theta} = \frac{1}{n}\cos\theta \tag{7.8}$$

$$\cos^2\phi\left(\frac{d\phi}{d\theta}\right)^2 = \frac{1}{n^2}\cos^2\theta \tag{7.9}$$

Using the trigonometric identity $\cos^2\phi + \sin^2\phi = 1$ we can therefore write:

$$(1 - \sin^2\phi)\left(\frac{d\phi}{d\theta}\right)^2 = \frac{1}{n^2}\cos^2\theta \tag{7.10}$$

and therefore again using Snell's Law:

$$\left(1 - \frac{1}{n^2}\sin^2\theta\right)\left(\frac{d\phi}{d\theta}\right)^2 = \frac{1}{n^2}\cos^2\theta$$

$$\left(\frac{d\phi}{d\theta}\right)^2 = \frac{\cos^2\theta}{n^2 - \sin^2\theta}$$

If $\frac{d\varepsilon}{d\theta} = 4\frac{d\phi}{d\theta} - 2$, the extremum of $\varepsilon(\theta)$ is when $\frac{d\varepsilon}{d\theta} = 0$, i.e. $4\frac{d\phi}{d\theta} - 2 = 0 \Rightarrow \left(\frac{d\phi}{d\theta}\right)^2 = \frac{1}{4}$.
Therefore:

$$\frac{\cos^2\theta}{n^2 - \sin^2\theta} = \tfrac{1}{4} \tag{7.11}$$

$$4\cos^2\theta = n^2 - \sin^2\theta$$

$$4\left(1 - \sin^2\theta\right) = n^2 - \sin^2\theta$$

$$4 - n^2 = 3\sin^2\theta$$

$$\therefore \theta = \sin^{-1}\sqrt{\frac{4 - n^2}{3}} \tag{7.12}$$

Now from Snell's law and the within-droplet ray tracing, the ray elevation is:

$$\varepsilon = 4\sin^{-1}\left(\frac{\sin\theta}{n}\right) - 2\theta \tag{7.13}$$

Hence the angle ε of the rainbow arc at the frequency of light corresponding to refractive index n is:

$$\varepsilon = 4\sin^{-1}\left(\sqrt{\frac{4-n^2}{3}} \times \frac{1}{n}\right) - 2\sin^{-1}\sqrt{\frac{4-n^2}{3}} \tag{7.14}$$

7.2.3 A double rainbow

Although a little more rare than the single variant, sightings of double rainbows (Fig. 7.3) are nonetheless commonplace, and certainly phenomena that can be easily observed with the naked eye. The secondary bow is at a larger elevation (about 51.6° from the anti-solar direction) than the primary bow (which is about 41.7°) and intriguingly the order of its coloured bands are *reversed*. Whereas a primary bow is purple, blue, green, yellow, orange, red at increasingly higher elevation angles, the opposite is true for the secondary bow. There is also the curious phenomenon of *Alexander's Dark Band*,[2] which is a darkening of the sky between the primary and secondary bows.

A very similar geometric construction to that of Fig. 7.2 can be made to explain the properties of secondary bow. In this case, light internally reflects *twice* in the interior of a spherical raindrop. As described in Fig. 7.3, the total deflection of the incident ray is:

$$\pi + \varepsilon = \underbrace{\theta - \phi}_{A} + \underbrace{\pi - 2\phi}_{B} + \underbrace{\pi - 2\phi}_{C} + \underbrace{\theta - \phi}_{D} \tag{7.15}$$

$$\therefore \varepsilon = \pi - 6\phi + 2\theta \tag{7.16}$$

where ε is the elevation of the ray exiting the raindrop to the observer, relative to the anti-solar direction (i.e. the incident ray from the sun). θ is the angle of incidence of the ray to the normal of the raindrop, and ϕ is the angle of refraction of the incident ray inside the raindrop.

---

[2] Alexander of Aphrodisias first described the effect in 200 AD. www.atoptics.co.uk/rainbows/adband.htm (accessed 8/1/2019).

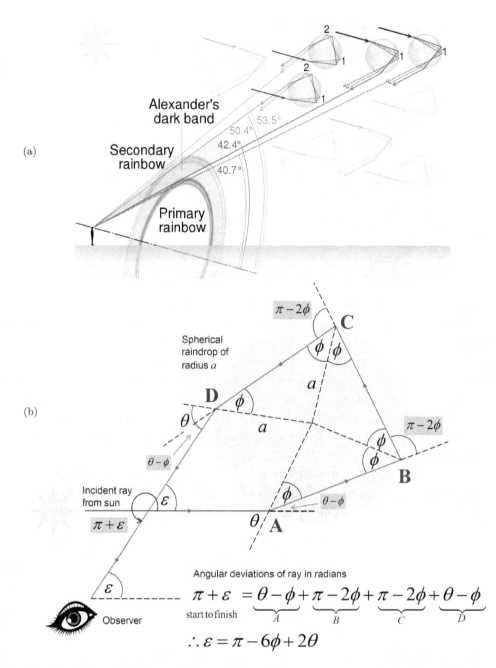

Fig. 7.3 Idealised model of a secondary rainbow. Light rays enter a spherical water droplet and are refracted by angle ϕ to the surface normal, as depicted in (b). These are internally reflected *twice* and then exit at an elevation angle ε. By tracing the ray path and use of Snell's law one can show $\varepsilon = \pi - 6\sin^{-1}\left(\frac{\sin\theta}{n}\right) + 2\theta$ where n is the refractive index. $\varepsilon(\theta)$ has a *minima*, which explains the formation of a rainbow at $\varepsilon = \pi - 6\sin^{-1}\left(\sqrt{\frac{9-n^2}{8}} \times \frac{1}{n}\right) + 2\sin^{-1}\sqrt{\frac{9-n^2}{8}}$, corresponding to the light colour associated with n. Physics of a primary and secondary rainbow in (a) is illustrated by CMG Lee. CC BY-SA 4.0. Wikipedia: Rainbow.

We can again make use of Snell's Law and its derivative with respect to ϕ, as shown in the previous section:

$$\sin \phi = \frac{\sin \theta}{n} \Rightarrow \left(\frac{d\phi}{d\theta}\right)^2 = \frac{\cos^2 \theta}{n^2 - \sin^2 \theta} \tag{7.17}$$

and use the idea that the double rainbow is at the elevation angle such that $\frac{d\varepsilon}{d\theta} = 0$. The elevation angle equation $\varepsilon = \pi - 6\phi + 2\theta$ implies:

$$\frac{d\varepsilon}{d\theta} = -6\frac{d\phi}{d\theta} + 2 \tag{7.18}$$

So if $\frac{d\varepsilon}{d\theta} = 0$, this means $-6\frac{d\phi}{d\theta} + 2 = 0 \Rightarrow \left(\frac{d\phi}{d\theta}\right)^2 = \frac{1}{9}$.

Therefore:

$$\frac{\cos^2 \theta}{n^2 - \sin^2 \theta} = \frac{1}{9} \tag{7.19}$$

$$9\cos^2 \theta = n^2 - \sin^2 \theta \tag{7.20}$$

$$9 - 9\sin^2 \theta = n^2 - \sin^2 \theta \tag{7.21}$$

$$9 - n^2 = 8\sin^2 \theta \tag{7.22}$$

$$\therefore \theta = \sin^{-1} \sqrt{\frac{9 - n^2}{8}} \tag{7.23}$$

$\varepsilon = \pi - 6\phi + 2\theta$ combined with $\theta = \sin^{-1} \sqrt{\frac{9-n^2}{8}}$ implies the elevation angle of the secondary bow is predicted to be:

$$\varepsilon = \pi - 6\sin^{-1} \left(\sqrt{\frac{9 - n^2}{8}} \times \frac{1}{n}\right) + 2\sin^{-1} \sqrt{\frac{9 - n^2}{8}} \tag{7.24}$$

7.2.4 *Variation of refractive index for water with frequency of light*

Hecht [21, p. 70], gives a formula for the frequency variation of the refractive index of a transparent medium. Using experimental data for the refractive index (of pure water at 20°C) we can deduce the following semi-empirical relationship between the refractive index of water at (optical) frequencies of light.

$$\left(n^2 - 1\right)^{-2} = \beta + \alpha \left(\frac{f}{10^{15}\text{Hz}}\right)^2 \tag{7.25}$$

where $\beta = 1.731$ and $\alpha = -0.261$. These numbers were calculated by performing a line of best fit of $y = \left(n^2 - 1\right)^{-2}$ vs $x = \left(\frac{f}{10^{15}\text{Hz}}\right)^2$ using measured values[3] for the refractive indices of light in pure water vs frequency. The intercept is β and the gradient α in this case.

---

[3]Refractive indices of pure water at 20°C. https://en.wikipedia.org/wiki/Optical_properties_of_water_and_ice.

The empirical formula above can be rearranged to make n the subject:

$$n = \sqrt{\left(\left(\beta + \alpha\left(\frac{f}{10^{15}\,\mathrm{Hz}}\right)^2\right)^{-\frac{1}{2}} + 1} \qquad (7.26)$$

Figure 7.4 is a plot of $y = \left(n^2 - 1\right)^{-2}$ vs $x = \left(\frac{f}{10^{15}\,\mathrm{Hz}}\right)^2$, and a standard linear-regression method has been used to find α, β. The measured data points (red crosses) are then underlaid with this parameterised model curve of n vs f.

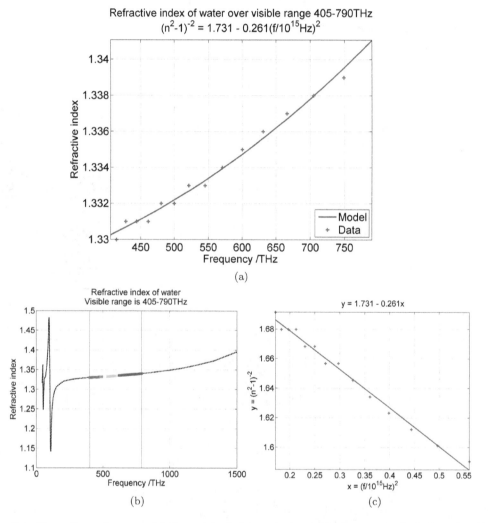

Fig. 7.4   Formation of an empirical model of the variation with frequency f of the refractive index n of water. This is done by fitting a suggested relationship $\left(n^2 - 1\right)^{-2} = \beta + \alpha\left(\frac{f}{10^{15}\,\mathrm{Hz}}\right)^2$ to measured data for water, and hence determining the parameters α and β from a line of best fit of $y = \left(n^2 - 1\right)^{-2}$ vs $x = \left(\frac{f}{10^{15}\,\mathrm{Hz}}\right)^2$. This yields the formula: $n = \sqrt{\left(1.731 - 0.261\left(\frac{f}{10^{15}\,\mathrm{Hz}}\right)^2\right)^{-\frac{1}{2}} + 1}$.

7.2.5 Using n vs frequency to determine the elevation of the primary and secondary rainbows

The empirical formula for refractive index n vs light frequency f can be used in conjunction with the elevation vs angle equation $\varepsilon(\theta)$ and the 'rainbow angle' equation; i.e. the angle θ_* when $\frac{d\varepsilon}{d\theta} = 0$.

$$n = \sqrt{\left(1.731 - 0.261\left(\frac{f}{10^{15}\text{Hz}}\right)^2\right)^{-\frac{1}{2}} + 1}$$

Primary rainbow

$$\varepsilon = 4\sin^{-1}\left(\frac{\sin\theta}{n}\right) - 2\theta \tag{7.27}$$

$$\theta_* = \sin^{-1}\sqrt{\frac{4 - n^2}{3}} \tag{7.28}$$

$$\varepsilon_* = \varepsilon(\theta_*) \tag{7.29}$$

Secondary rainbow

$$\varepsilon = \pi - 6\sin^{-1}\left(\frac{\sin\theta}{n}\right) + 2\theta \tag{7.30}$$

$$\theta_* = \sin^{-1}\sqrt{\frac{9 - n^2}{8}} \tag{7.31}$$

$$\varepsilon_* = \varepsilon(\theta_*) \tag{7.32}$$

Figure 7.5(a) plots the bow elevation ε vs incident angle θ. The colours of the curves represent the visible colours associated with a particular frequency and hence refractive index n. Curves are overlaid for 100 linearly spaced frequencies between 405 THz and 790 THz. The primary bow curve has a maximum at around $\varepsilon = 41.7°$, whereas the secondary bow has a minimum at around $\varepsilon = 51.6°$. The coloured horizontal lines represent the small deviations in these stationary points for the different frequencies, and hence the range of elevations quoted. (For example 50.2° to 53.0° for the secondary bow.) The variation of bow elevation angles are more clearly described in Fig. 7.5(b), in which elevation angle ε_* of the rainbow is plotted against frequency f.

In both these figures, the reversal of the colour orders is clear. In terms of the light frequency (and hence energy inherent in the light[4]), the higher frequency light is at higher elevations for the secondary bow, and lower elevations for the primary bow. Therefore one might expect a reduction in light intensity between the rainbows, which explains *Alexander's Dark Band*.

---

[4]The energy E of a photon of light of frequency f is given by $E = hf$, where Planck's constant $h = 6.63 \times 10^{-34}\,\text{m}^2\,\text{kgs}^{-1}$.

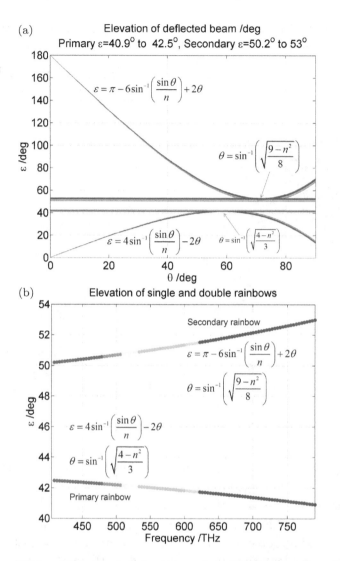

Fig. 7.5 (a) The variation of the observed elevation angle ε of internally refracted light rays which are at incident angle θ from the normal to a raindrop are plotted for different colours (and hence different refractive indices) and for one (primary rainbow) and two (secondary rainbow) internal reflections. The horizontal lines correspond to the elevation angles of the *extremum* (i.e. maxima or minima) of the $\varepsilon(\theta)$ curves. These are the angles of the rainbow: about 40.9° to 42.5° for a primary rainbow, and 50.2° to 53° for a secondary bow. Note these angles are relative to the anti-solar direction, so if the sun is tilted at a large angle α, then one may only see a minor arc of the rainbow. (b) Elevation angles of primary and secondary rainbows are plotted against optical light frequencies, and suitably colour coded.

7.2.6 *Rainbow arcs and the angle of the anti-solar direction*

The idea that a rainbow is formed at a particular elevation angle from the anti-solar direction (i.e. parallel rays from the sun, incident on rain droplets) implies a cone of light with the observer at the apex. If the light source is directly behind the observer, and there are no topographic obstructions, one should see a complete circle

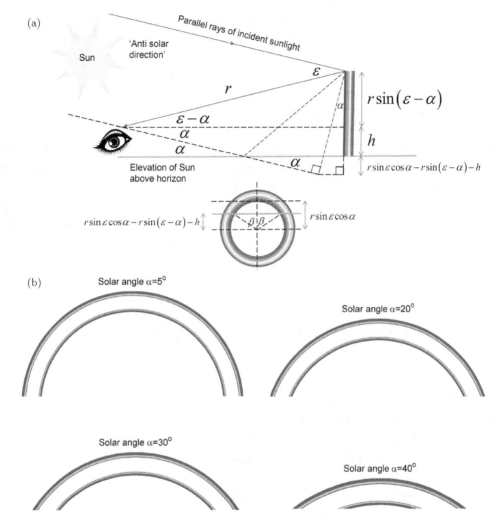

Fig. 7.6 (a) Model of primary of primary and secondary rainbow arcs as seen by an observer at height h, given the Sun, in the anti-solar position, at elevation angle α. For the arcs drawn in (b), $h = 0$. One can clearly see why rainbows appear larger when the Sun is low.

rather than the more familiar arc. This is indeed how a rainbow might be observed from the flight deck of an aircraft. On the Earth's surface, the horizon will be the ultimate limit to rainbow visibility. Also, the anti-solar direction will rise behind the observer as the Sun does. Once this angle rises to beyond $\varepsilon = 41.7°$, there will be no visible primary rainbow, and once beyond $51.6°$, no secondary rainbows either. As indicated in Fig. 7.6, rainbows are best observed when the Sun is low, i.e. during the early morning and evening.

7.2.7 *Higher-order rainbows*

The critical angle for water is approximately $\sin^{-1}\left(\frac{1}{1.33}\right) \approx 48.8°$. Not all values of the angle ϕ will be greater than this, so there will be some transmission through the raindrop as well as internal reflection. We would therefore anticipate a loss in intensity of light with every internal reflection, and this helps to explain why higher order bows don't tend to be seen with the naked eye. (As mentioned at the start of this chapter, recent observations have required the use of powerful image processing techniques to reveal higher order bows.) Also, the geometry of repeated internal reflections inside a sphere will preclude certain ray paths from exiting the raindrop so that they can be observed from below, with the Sun behind you. This explains why you don't see tertiary rainbows at some higher elevation than the 53.0° of secondary rainbows.[5] However, from a ray-tracing perspective it is possible to generate much higher order bows, particularly if highly spherical water droplets can be illuminated by a laser. In the *Bending Light* simulation from the University of Colorado's most excellent online educational PhET resource,[6] I have personally counted over thirteen internal reflections if the incident ray is very close to a tangent to the spherical water droplet.

7.3 Coda

I have righteously lambasted the artsy folk in this chapter with a reductionist rainbow riposte. However, I do rather like verse. So for balance, I offer the following poetic pair:

Sonnet (sonnet)
Thirteenth-century Tuscan poets proposed
(Namely Petrarch, Lentini and Dante)
That only fourteen lines need be composed
Else a verse is too long, or too scanty
John Milton's blank verse was a paradise lost
As he filled ten books with his ode
None fretted about the narrative cost
For each word's worth a drop in this load
And what of Shakespeare, Auden, Owen and Keats?
Hard rime on a dull English shackle?
Although youth is doomed to recite complete
No more than twice-seven lines need be tackled
But change of units is ictic, under poesy's bonnet...
As metres and feet, are both used, in a sonnet!

---

[5]You *might* be able to see a tertiary rainbow if you were to look (carefully!) in the direction of the Sun, rather than in the anti-solar direction, as assumed in the rainbow model of this chapter.
[6]https://phet.colorado.edu/en/simulation/bending-light.

Morin, the Classical Mechanic

If strings, branes and quantum foams
Aren't easily digested
There is a book (it's coloured red)
That time is well invested
If balls and springs aren't your thing
Your Physics inclination planar
Try sliding rings on inextensible things
And force yourself to be saner
Gravity acts in silent space
There's no torque but plenty of action
Orbits are stable, elliptically
Since periods aren't rational fractions
Cease being fooled, try Lagrangian tools
If you are chained and under the cosh
It's Noether here nor there, if constraints are there
Don't despair, your equations will wash!
And if $\mathbf{r} \times \mathbf{p}$ isn't momentous for you
And your list of problems ain't topped
Feast your eyes on Relativity too
And dilate your pupils via clocks
And for those, who desire yet more
Go to unexpected lengths with vectors-four
In summary we all like this tome
It's even on my desk at home
For a Physics fix, no need to panic . . .
Try Morin, the Classical Mechanic

The latter was written as part of an entirely geeky *Morinfest* arranged for my Physics sixth form class of 2017–2018. David Morin, a lecturer in Physics at Harvard University, wrote the tour-de-force which is his big red book of Classical Mechanics problems and clever limericks [29]. All problems are stated with beautiful simplicity, but often require serious effort to solve. Certain members of my class were obsessed with the book, and its * rating for problem difficulty. (* a challenge, ** eye-wateringly hard, *** teacher spends the weekend doing it, **** research project.) The enthusiasm for problem solving was too good to let dissipate, so I arranged an evening where each student could present their favourite solution, with the proviso that they also compose their own Morin-style limerick. Of course it was important that their teacher should have a go too. . .

Chapter 8

Exploring Julia's Fractals

8.1 The Fractal Geometry of Nature

Fractals are infinitely complex geometric objects that have the strange property of looking very similar at all scales. Many structures in nature are like this. The jagged shape of a mountain range resembles the rocks that form it, and if you look at a rock under a powerful microscope you will see the jagged structure repeat, until you reach atomic dimensions. Amazingly, it is possible to create fractal patterns that are based upon very simple mathematical rules, applied recursively, i.e. in loops, or *iterations*. We shall look at the beautiful structures (e.g. in Fig. 8.3) discovered by Gaston Julia (1893–1978) and re-discovered and popularised by Benoit Mandelbrot (1924–2010). The work of Mandelbrot, particularly following the publication of his classic book *The Fractal Geometry of Nature* [28], has inspired thousands of programmers to write software to re-create the images he revealed to be associated with a very simple quadratic equation. One of the great joys of the Mandelbrot fractals is that the code to create them is very simple, and pretty much any personal computer since the 1980s has the power to render them. My own personal programming journey over the past 20 years really began with the creation of `julia`, which is a program for generating large format generalisations of Mandelbrot fractals, using a Graphical User Interface (GUI) coded in the MATLAB environment.

8.2 Mandelbrot Fractals

The Mandelbrot fractals are based upon an iterative transformation of *complex numbers* (see the following section). An array of complex numbers z_0, with *real* and *imaginary* parts over the range $[-\pi, \pi]$ is modified iteratively using the transformation:

$$z \rightarrow f\left(z^2 + z_0\right) \tag{8.1}$$

147

where z is initially z_0, as illustrated in Figs. 8.1 and 8.2. Two types of image are created after between twenty and thirty iterations of the transformation. The classic *Mandelbrot* fractal is obtained by determining whether the magnitude $|z|$ of each point in the array of complex numbers exceeds a defined value (e.g. 5) during each iteration. When this happens, the *iteration number* is associated with the location in the complex plane defined by the original array z_0. After a specified maximum number of iterations has been computed, a map of the iteration numbers is plotted. In Fig. 8.3, a white colour represents the set of array elements which have not yet diverged in magnitude from the defined maximum of $|z|$, the 'radius of convergence'. These are said to be '*within the Mandelbrot Set*'. Intriguingly, even for the most simple transformation $z \to z^2 + z_0$, this represents a non-obvious shape. Inspection of its boundary reveals an *infinite* level of complexity and, if one zooms in on particular regions, a repetition of the original figure, i.e. the fractal property of *self-similarity* (see Fig. 8.3).

More sophisticated versions of $f(z)$ produce distinct and interesting shapes. A different mechanism is used to render most of these. Instead of plotting the iteration number when $|z|$ exceeds a given value, the final complex number array z (i.e. after N iterations) is plotted following a complex to real number transformation such as $z \to \arg(z)$ or $z \to e^{-|z|}$. In this case it is the array elements whose magnitude *does not diverge* which contribute to the most interesting structures. For artistic purposes a prismatic colour scheme is chosen for this chapter's figures, which cycles through its colour spectrum several times over the range of the resulting plot map magnitudes.

The trajectory of complex numbers starting with a particular z_0 is illustrated in Figs. 8.1 and 8.2. For Fig. 8.2, a coloured surface of $e^{-|z|}$ is plotted, and clearly shows the beetle-like multi-cardioid shape of the Mandelbrot fractal. Figure 8.2 illustrates two screenshots of a dynamic calculation of 50 classic Mandelbrot iterations (i.e. $z_{n+1} = z_n^2 + z_0$) starting from a z_0 value which is set by the current position of the mouse pointer. Around the border region, a wild and complicated variety of shapes are created, from spirals, to multi-pointed stars, and indeed the type of rotated triangles that were associated with the *Snails of Pursuit* in the introductory chapter! The trajectories inspiral ever closer as one gets closer to $(0,0)$, and diverge when one is beyond the Mandelbrot border region. This is why the border region of the Mandelbrot fractal yields the most intricate, and most interesting geometries.

The majority of the images in this chapter are generated using a bespoke MATLAB application `julia`, which was created by the author between 2004 and 2011 (see Figs. 8.3 and 8.4). The name pays homage to Gaston Julia (1893–1978), whose work in 1918 on the *Iteration of Rational Functions* was incorporated into the theory of *fractal geometry* developed by Benoit Mandelbrot and popularised in his highly influential book *The Fractal Geometry of Nature* (1982). A selection

(a) Mandelbrot transformations of **complex numbers**

$$i^2 = -1$$

$$z = x + iy$$

$$x = \mathrm{Re}(z)$$

$$y = \mathrm{Im}(z)$$

$$|z| = \sqrt{x^2 + y^2}$$

$$(1+i)(1+i)$$

$$= 1 + 2i + i^2$$

$$= 1 + 2i - 1$$

$$= 2i$$

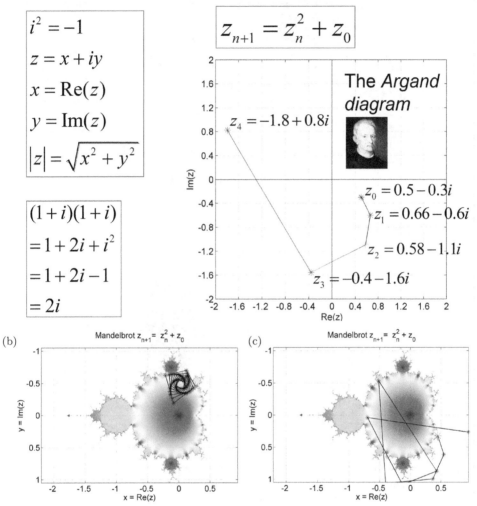

$$z_{n+1} = z_n^2 + z_0$$

The *Argand diagram*

$z_4 = -1.8 + 0.8i$

$z_0 = 0.5 - 0.3i$

$z_1 = 0.66 - 0.6i$

$z_2 = 0.58 - 1.1i$

$z_3 = -0.4 - 1.6i$

(b) Mandelbrot $z_{n+1} = z_n^2 + z_0$

(c) Mandelbrot $z_{n+1} = z_n^2 + z_0$

Fig. 8.1 *Complex numbers* are an *imaginary* extension of the traditional *real* number line extending horizontally from increasingly negative numbers to increasingly positive. A number with an imaginary (vertical) component is a real number y multiplied by i, which means a unit along the (vertical) imaginary axis. Complex numbers of the form $z = x + iy$ can be plotted on a Cartesian grid called an *Argand Diagram*. Complex numbers add and multiply just like real numbers, with the extra rule that $i^2 = -1$. A complex iteration $z_{n+1} = z_n^2 + z_0$ produces quite startling complexity. An example iteration is plotted in (a). (b) shows an iteration starting with z_0 within the 'Mandelbrot set', a complex beetle-like shape, that as underlaid in colour for reference. The closer z_0 is to the boundary, the more intricate the trajectory becomes. (c) starts with z_0 outside the Mandelbrot set, and this trajectory rapidly diverges.

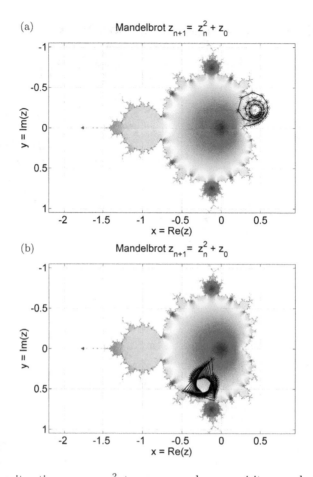

(a) Mandelbrot $z_{n+1} = z_n^2 + z_0$

(b) Mandelbrot $z_{n+1} = z_n^2 + z_0$

Fig. 8.2   A complex iteration $z_{n+1} = z_n^2 + z_0$ can produce exquisite complexity. Example iterations are plotted in subplots (a) and (b). Both illustrate an iteration starting with z_0 within the 'Mandelbrot set', a complex beetle-like shape, that as underlaid in colour for reference. The closer z_0 is to the boundary, the more intricate the trajectory becomes. A huge range of interesting trajectories are possible, from spirals (a) to rotated triangles (b). Both images are screenshots from a MATLAB computer program. The first 50 points of the iteration $z_{n+1} = z_n^2 + z_0$ are plotted based upon a z_0 value fixed by the current location of the mouse pointer. The user can move the mouse and hence explore the convergence or divergence of the iteration dynamically.

of interesting images, which I have assembled as *The Mandelbrot Variations* (see Fig. 8.5) can be created via the following transformations: $f(z) = z$, $f(z) = \sin z$, $f(z) = \sin^{-1} z$, $f(z) = \tan^{-1} z$, $f(z) = z^2$, $f(z) = z^*$, $f(z) = \log z$, $f(z) = 1/z$, $f(z) = z^z$. The $f(z) = z^2$ transformation (top left) produces an image that could easily be an image of a micro-organism. If viewed on its side, the $f(z) = \tan^{-1} z$ (one down from top left) might look like a piece of Buddhist art representing the 'eightfold path to enlightenment'. (There are about eight prismatic steps, if you use your imagination!) The $f(z) = \log z$ variation (middle, second from bottom)

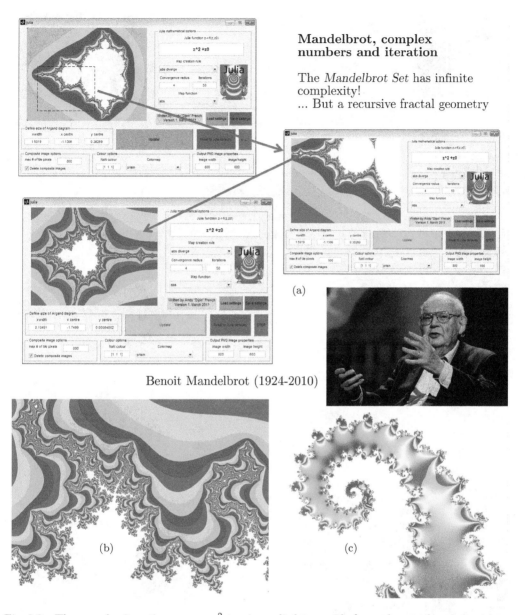

Mandelbrot, complex numbers and iteration

The *Mandelbrot Set* has infinite complexity!
... But a recursive fractal geometry

(a)

Benoit Mandelbrot (1924-2010)

(b)

(c)

Fig. 8.3   The complex iteration $z_{n+1} = z_n^2 + z_0$ is applied to a grid of complex numbers in matrix z_0. When $|z| > 4$, the iteration number is logged for the particular complex number in z_0 that satisfies this criteria, and then this number is ignored from subsequent iterations to help speed up the process. (a) The MATLAB app `julia` plots the resulting map of iteration numbers using a prismatic colour scale. Zooming in and regenerating this *Mandelbrot Set* reveals an infinite, organic complexity, with plant-like structures (b) and spiral tails (c). Benoit Mandelbrot image (cropped). Original by Steve Jurvetson. CC by 2.0. Wikipedia: Benoit Mandelbrot.

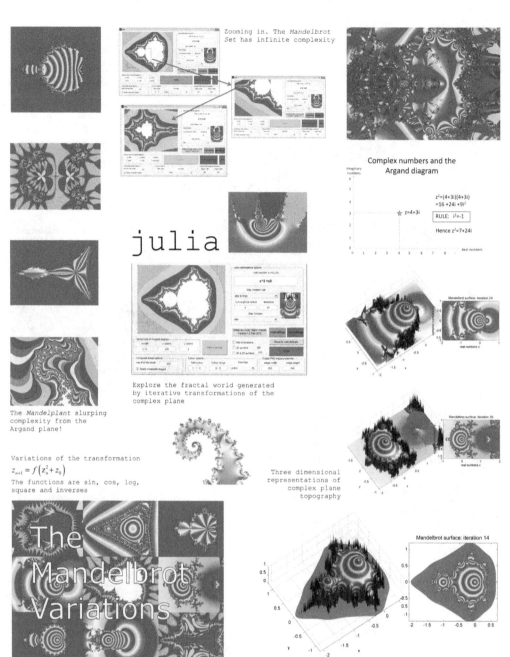

Fig. 8.4   A poster showcasing the outputs of the MATLAB computer program `julia`. The *Mandelbrot Variations* are shown, and indeed the three-dimensional surfaces whose height (above the complex plane) relates to a (real-valued) function of the final state of the complex plane after a number of Mandelbrot-style iterative transformations.

Fig. 8.5 Many variations upon the Mandelbrot set visualisation recipe. ** implies the colour scheme is based upon a scalar function (e.g. $h = \arg(z)$ or $e^{-|z|}$) of the $z = x + iy$ coordinates of the Argand diagram after a certain number of iterations. Otherwise, assume the classic Mandelbrot recipe i.e. the colour scale relates to the number of iterations when an initial point $z_0 = x + iy$ diverges from a specified radius of convergence, and is therefore outside the Mandelbrot Set. Scanning right from top left: (i) $z_{n+1} = (z_n^2 + z_0)^2$, (ii)** $z_{n+1} = (z_n^2 + z_0)^{-1}$, (iii)** $z_{n+1} = z_n^2 + z_0$, $h = \arg(z)$, (iv)** $z_{n+1} = \tan^{-1}(z_n^2 + z_0)$, (v) iteration number plot where both magnitudes of real *and* imaginary parts of z exceed convergence radius, (vi)** $z_{n+1} = \sin^{-1}(z_n^2 + z_0)$, (vii)** $z_{n+1} = z_n^2 + z_0$, $h = \sin(|z|)$, (viii)** $z_{n+1} = \ln(z_n^2 + z_0)$, (ix) $z_{n+1} = (z_n^2 + z_0)^*$, (x) $z_{n+1} = (z_n^2 + z_0)^{-1} + z_n^2 + z_0$, (xi)** $z_{n+1} = (z_n^2 + z_0)^*$, $h = \arg(z)$, (xii)** $z_{n+1} = (z_n^2 + z_0)^{z_n}$.

looks like light radiating out from a filament bulb, and the $f(z) = \sin^{-1} z$ variation resembles a strange organic rocket-ship. The $f(z) = z^*$ variant (with a $z \to \arg(z)$ colour display function), bottom, middle, resembles a supersonic aircraft. The bottom left image, formed from the summation of the Mandelbrot transformation and its inverse, i.e. $f(z) = z + \frac{1}{z}$ is strangely my favourite, for reasons I can't really rationalise. It is on my study wall, top left of where I am typing this now.

It is worth emphasising that the images formed by plotting a coloured surface of height $e^{-|z|}$, or $\arg(z)$, are actually three-dimensional, but simply viewed from above in *The Mandelbrot Variations*. The first 24 iterations of the classic Mandelbrot $z_{n+1} = z_n^2 + z_0$ are plotted in an art piece I call *The Day of Julia*, as each iteration represents an hour (see Fig. 8.6). The progression starts as an exotic tent, which then begins to resemble the great Mosque of *Hagia Sophia* in Istanbul, resplendent with domes and minnarets. Towards the end of the 'day', the 'minnaretia' becomes a dense boundary of spines around a smooth cone-like covering of a cardioid shape. At this point the three-dimensional structure fails to convey the complexity, which is now confined to the boundary region. From an aesthetic perspective, a two-dimensional visualisation now offers the greater riches.

A popular test of programming prowess is to perform a 'Mandelbrot dive', which is a video of a deep zoom into the Mandelbrot fractal. An impressive (but certainly not record breaking) example reports an extraordinary zoom factor of 2.1×10^{275}, and the accompanying video[1] and pulsating soundtrack may well be considered a psychedelic experience. The Mandelbrot dive works by successively recalculating a Mandelbrot image using a z_0 grid of complex numbers from an x, y range which is increasingly narrow. `julia` is not highly optimised by modern coding standards, but it does incorporate a few speed-boosting tricks. Firstly the use of vector processing, which means the Mandelbrot iterations are applied to all complex numbers in the grid which represents the Argand diagram simultaneously, rather than one at a time. Without vector processing, one would require a double `for` loop to work out the transformations of every element of the Argand diagram, for each iteration, and this *serial* process would be significantly slower from a computational perspective. In addition, z_0 elements that eventually diverge can be ignored from subsequent iterations, which reduces the computational burden in subsequent recursions. However, in spite of these tricks, it will still take a moderately powerful personal computer several seconds to render just one frame at HD resolution (i.e. 1920×1080 pixels), so to create a 10-minute movie of a deep Mandelbrot zoom requires serious computation. An additional subtlety is the ability to deal with increasingly large numbers of significant figures, so the Mandelbrot calculator must be able to incorporate arbitrary precision arithmetic too. `julia` is not that sophisticated, and works with the default limits of numerical precision in MATLAB. For 64-bit systems, the smallest number is $2.225073858507201 \times 10^{-308} = 2^{-1022}$.

---

[1] Deep Mandelbrot zoom: https://www.youtube.com/watch?v=0jGaio87u3A.

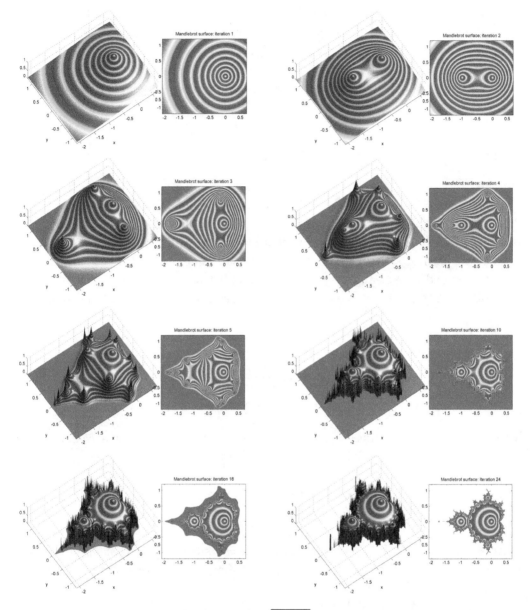

Fig. 8.6 A surface of height $h_n(x,y) = \exp(-\sqrt{x^2 + y^2})$ is plotted in two and three dimensions for iteration n over range $n = 1, 2, 3, \ldots, 24$. A grid of complex numbers z_0 are iterated via $z_{n+1} = z_n^2 + z_0$ and $x = \text{Re}(z_n)$ and $y = \text{Im}(z_n)$. As n increases, the density of boundary pinnacles around the circles and cardioid bean-like structure increases significantly.

8.3 Complex Numbers and de-Moivre's Theorem

8.3.1 *Real and imaginary numbers*

Complex numbers are a geometric extension to the number system. If a *real* number like 1, -4, $\frac{1}{3}$, $\sqrt{2}$, $-\pi \ldots$ can be represented as a point on a horizontal number line, a *complex number* has an additional *vertical* part. This is known as the *imaginary axis*. A general number z can therefore be represented as an (x, y) coordinate in the complex plane, or *Argand diagram* (see Fig. 8.1).

The fundamental difference between a complex number and a two-dimensional *vector* is the algebraic relationship between the real and imaginary parts of z:

$$z = x + iy \tag{8.2}$$

where the 'imaginary number' i is defined to satisfy:

$$i^2 = -1 \tag{8.3}$$

i.e. z is a summation of *scalar* numbers. A *vector* addition, such as $\mathbf{r} = x\widehat{\mathbf{x}} + y\widehat{\mathbf{y}}$, is different, because x, y scale the *unit vectors* $\widehat{\mathbf{x}}, \widehat{\mathbf{y}}$ which represent the mutually orthogonal axes of a Cartesian system. When it comes to multiplication, x and y are not blended in the same way. Complex number multiplication is just like scalar multiplication, but with $i^2 = -1$ allowing for a subtle mixing of real and imaginary parts.

$$z_1 z_2 = (x_1 + iy_1)(x_2 + iy_2) \tag{8.4}$$

$$= x_1 x_2 + i(x_1 y_2 + x_2 y_1) + i^2 y_1 y_2 \tag{8.5}$$

$$= x_1 x_2 - y_1 y_2 + i(x_1 y_2 + x_2 y_1) \tag{8.6}$$

Vector multiplication proceeds either as a *scalar product* $\mathbf{r}_1 \cdot \mathbf{r}_2$ or a *vector product* $\mathbf{r}_1 \times \mathbf{r}_2$. For two-dimensional vectors $\mathbf{r} = x\widehat{\mathbf{x}} + y\widehat{\mathbf{y}}$:

$$\mathbf{r}_1 \cdot \mathbf{r}_2 = (x_1\widehat{\mathbf{x}} + y_1\widehat{\mathbf{y}}) \cdot (x_2\widehat{\mathbf{x}} + y_2\widehat{\mathbf{y}}) \tag{8.7}$$

$$= x_1 x_2 + y_1 y_2 \tag{8.8}$$

This is because $\widehat{\mathbf{x}} \cdot \widehat{\mathbf{x}} = 1$, $\widehat{\mathbf{y}} \cdot \widehat{\mathbf{y}} = 1$ and $\widehat{\mathbf{x}} \cdot \widehat{\mathbf{y}} = 0$ since $\widehat{\mathbf{x}}, \widehat{\mathbf{y}}$ are *orthonormal*. (i.e. perpendicular to each other, and both of unit magnitude). For the cross-product:

$$\mathbf{r}_1 \times \mathbf{r}_2 = (x_1\widehat{\mathbf{x}} + y_1\widehat{\mathbf{y}}) \times (x_2\widehat{\mathbf{x}} + y_2\widehat{\mathbf{y}}) \tag{8.9}$$

$$= x_1 x_2 \widehat{\mathbf{x}} \times \widehat{\mathbf{x}} + y_1 x_2 \widehat{\mathbf{y}} \times \widehat{\mathbf{x}} + x_1 y_2 \widehat{\mathbf{x}} \times \widehat{\mathbf{y}} + y_1 y_2 \widehat{\mathbf{y}} \times \widehat{\mathbf{y}} \tag{8.10}$$

$$= (x_1 y_2 - y_1 x_2)\widehat{\mathbf{x}} \times \widehat{\mathbf{y}} \tag{8.11}$$

$$= (x_1 y_2 - y_1 x_2)\widehat{\mathbf{z}} \tag{8.12}$$

Note $\widehat{\mathbf{x}} \times \widehat{\mathbf{x}} = \widehat{\mathbf{y}} \times \widehat{\mathbf{y}} = 0$, $\widehat{\mathbf{y}} \times \widehat{\mathbf{x}} = -\widehat{\mathbf{x}} \times \widehat{\mathbf{y}}$, and since Cartesian unit vectors form a right-handed set: $\widehat{\mathbf{x}} \times \widehat{\mathbf{y}} = \widehat{\mathbf{z}}$. The evaluation of the vector product requires an additional $\widehat{\mathbf{z}}$ axis that is mutually perpendicular to both $\widehat{\mathbf{x}}$ and $\widehat{\mathbf{y}}$, so in this case we cannot entirely constrain ourselves to two dimensions.

i.e. the distance of (x, y) from the origin is r and the angle (in radians) measured anticlockwise from the horizontal axis is θ. In complex number terminology, $r = |z|$ is the absolute value or magnitude of z, whereas $\theta = \arg(z)$ is the argument (or perhaps, subject to a minus-sign convention, 'phase') of z.

Hence by de-Moivre's theorem:

$$z = x + iy = re^{i\theta} \tag{8.16}$$

A beautiful relationship can be formed, if one starts with the definition of i as $y = 1$, $x = 0$, which also means $r = 1, \theta = \frac{\pi}{2}$.

$$i = e^{i\frac{\pi}{2}} \tag{8.17}$$

$$i^2 = e^{i\pi} = -1 \tag{8.18}$$

So "e to the i pi's a minus,"[5] or even better:

$$e^{i\pi} + 1 = 0 \tag{8.19}$$

This is known as the *Euler relation*, in homage to the great Swiss mathematician Leonhard Euler (1707–1783). All the basic numbers, $0, 1, i, e, \pi$ are connected via a typographic arrangement of the basic symbols $+$, $=$, exponentiation and multiplication. If a distillation of mathematics is required, this is it. 100% proof in every sense! It ought to be printed upon all future space probes that leave the solar system. de-Moivre's theorem means we can make use of powerful simplifications when evaluating *powers* of z, since:

$$z^n = (x + iy)^n = r^n e^{i\theta n} \tag{8.20}$$

For example: $(\cos\theta + i\sin\theta)^{99} = \cos(99\theta) + i\sin(99\theta)$. This may sound like a pretty niche result, but the idea behind it is very useful in working out trigonometric identities such as:

$$\sin^3\theta = \tfrac{1}{4}(3\sin\theta - \sin 3\theta) \tag{8.21}$$

The trick is to use $z = e^{i\theta} = \cos\theta + i\sin\theta$ which means:

$$2\cos\theta = z + \frac{1}{z} \tag{8.22}$$

$$2i\sin\theta = z - \frac{1}{z} \tag{8.23}$$

---

[5]Yes, there is a song with this as the lyric. I have based it (very loosely) on the Beatles' classic *Lucy in the Sky with Diamonds*. http://www.eclecticon.info/music_yours_truly_science.htm.

and hence:

$$2\cos n\theta = z^n + \frac{1}{z^n} \tag{8.24}$$

$$2i\sin n\theta = z^n - \frac{1}{z^n} \tag{8.25}$$

Using sine as an example, the idea is to raise $2i\sin\theta$ to a desired (positive) integer power n, expand out $\left(z - \frac{1}{z}\right)^n$ binomially, and then collect terms in $z^m - \frac{1}{z^m}$ where m is an integer $\leq n$. Then write this in terms of $\sin m\theta$, for all the values of m.

$$(2i\sin\theta)^3 = \left(z - \frac{1}{z}\right)^3 = z^3 + 3z^2\left(-\frac{1}{z}\right) + 3z\left(-\frac{1}{z}\right)^2 + \left(-\frac{1}{z}\right)^3 \tag{8.26}$$

$$= z^3 - 3z + \frac{3}{z} - \frac{1}{z^3} \tag{8.27}$$

$$= z^3 - \frac{1}{z^3} - 3\left(z - \frac{1}{z}\right) \tag{8.28}$$

$$= 2i\sin 3\theta - 3\left(2i\sin\theta\right) \tag{8.29}$$

$$-8i\sin^3\theta = 2i\sin 3\theta - 6i\sin\theta \tag{8.30}$$

After dividing both sides of the equation by $-8i$ and noting $1/i = -i$, we arrive at the required trigonometric identity:

$$\sin^3\theta = \tfrac{1}{4}\left(3\sin\theta - \sin 3\theta\right) \tag{8.31}$$

Also, since polar angle rotations of integer N multiples of 2π yield the same complex number, we can more generally interpret the nth root of a (complex) number:

$$z^{1/n} = (x + iy)^{1/n} = \sqrt[n]{r}\,e^{i\left(\frac{\theta}{n} + 2\pi\frac{N}{n}\right)} \tag{8.32}$$

For example, the nth *root of unity* is the solution to $z^n = 1 = e^{2\pi i N}$, where n is a positive integer, and N is an integer from the set $1...n$. Hence:

$$z = \omega, \omega^2, \ldots, \omega^n \tag{8.33}$$

where

$$\omega = e^{2\pi i/n} \tag{8.34}$$

i.e. in the Argand diagram, the nth root of unity form the vertices of an n sided regular polygon, with 'radials' (i.e. the distance to every vertex from the origin) being unity. In a similar way, logarithms of complex numbers are also multi-valued, and can be evaluated using de-Moivre's theorem:

$$z = re^{i(\theta + 2\pi N)} \tag{8.35}$$

$$\ln z = \ln r + i\left(\theta + 2\pi N\right) \tag{8.36}$$

For example: $3 + 4i = \sqrt{3^2 + 4^2}e^{i\left(\tan^{-1}\frac{4}{3}+2\pi N\right)}$ means:

$$\log(3 + 4i) = \log 5 + i \tan^{-1}\tfrac{4}{3} + 2\pi i N \tag{8.37}$$

8.4 Proof of de-Moivre's Theorem

Probably the most popular proof of de-Moivre's theorem is via a *Maclaurin expansion* (where $f^{(n)}(x) = \frac{d^n f(x)}{dx^n}$, i.e. the nth derivative):

$$f(x) = f(0) + x f'(0) + \frac{x^2 f''(0)}{2!} + \frac{x^3 f'''(0)}{3!} + \cdots + \frac{x^n f^{(n)}(0)}{n!} + \cdots \tag{8.38}$$

The Maclaurin expansions of $\sin\theta$, $\cos\theta$ and $e^{i\theta}$ are (assuming θ is in radians):

$$\cos\theta = 1 - \frac{\theta^2}{2!} + \frac{\theta^4}{4!} - \frac{\theta^6}{6!} + \frac{\theta^8}{8!} - \cdots \tag{8.39}$$

$$\sin\theta = \theta - \frac{\theta^3}{3!} + \frac{\theta^5}{5!} - \frac{\theta^7}{7!} + \frac{\theta^9}{9!} - \cdots \tag{8.40}$$

$$e^{i\theta} = 1 + (i\theta) + \frac{(i\theta)^2}{2!} + \frac{(i\theta)^3}{3!} + \frac{(i\theta)^4}{4!} + \frac{(i\theta)^5}{5!} \cdots \tag{8.41}$$

Hence using $i^2 = -1$:

$$e^{i\theta} = 1 - \frac{\theta^2}{2!} + \frac{\theta^4}{4!} - \frac{\theta^6}{6!} + \frac{\theta^8}{8!} - \cdots + i\left(\theta - \frac{\theta^3}{3!} + \frac{\theta^5}{5!} - \frac{\theta^7}{7!} + \frac{\theta^9}{9!} - \cdots\right) \tag{8.42}$$

$$e^{i\theta} = \cos\theta + i\sin\theta \tag{8.43}$$

A possibly more elegant proof uses calculus ideas directly, and also makes good use of $i^2 = 1$. Consider a complex number of unit magnitude, i.e. $|z| = 1$.

$$z = \cos\theta + i\sin\theta \tag{8.44}$$

$$iz = i\cos\theta - \sin\theta \tag{8.45}$$

$$\frac{dz}{d\theta} = -\sin\theta + i\cos\theta \tag{8.46}$$

Hence:

$$\frac{dz}{d\theta} = iz \tag{8.47}$$

A general solution to this differential equation is $z = Ae^{i\theta}$, which is clearly seen by noting $\frac{d}{d\theta}(Ae^{i\theta}) = iAe^{i\theta} = iz$. Now since $|z| = 1$, this means $A = 1$. Hence:

$$e^{i\theta} = \cos\theta + i\sin\theta \tag{8.48}$$

8.5 Making It Complex Can Make It Simpler

This chapter is basically a praise piece for the mathematical wonder-idea of complex numbers. A myriad of tedious algebraic operations can be made elegant by using complex numbers, and the fractal beauty and mindboggling complexity of the *Mandelbrot Set* can surely only be a learning incentive, particularly as it is so simple to code in software.

Much of physics involves the modelling of *wave phenomena*, and using Fourier's ideas,[6] we can imagine any wave-like disturbance to be the *sum of cosine or sine waves*, with different frequencies, amplitudes and phase shifts. The most basic waveform *Fourier component* is $\psi(x,t) = A\cos(2\pi\frac{x-ct}{\lambda} - \phi) = A\cos(kx - \omega t - \phi)$ where c is the wave speed, λ is the wavelength and ϕ is a phase shift. x, t are, respectfully, displacement and time. *Wavenumber* $k = \frac{2\pi}{\lambda}$ and *angular frequency* $\omega = \frac{2\pi c}{\lambda} = 2\pi f$ where f is the frequency. It is often useful to work out derivatives of wave disturbance ψ, as this might correspond to a particle velocity, and in the case of Schrödinger's famous equation[7] of Quantum Mechanics, a second derivative is present in the first term. Sine and Cosine terms interchange in a cumbersome way when calculus operations are applied, whereas exponentials are much easier to deal with. $\frac{d}{dx}e^{ax} = ae^{ax}$, whereas $\frac{d}{dx}\cos(ax) = -a\sin(ax)$ and $\frac{d}{dx}\sin(ax) = a\cos(ax)$. Hence when dealing with wave phenomena, we note:

$$\text{Re}\left(Ae^{i(kx-\omega t-\phi)}\right) = A\cos(kx - \omega t - \phi) \tag{8.49}$$

and hence use:

$$\psi(x,t) = Ae^{i(kx-\omega t-\phi)} \tag{8.50}$$

as our wave disturbance.

---

[6] Joseph Fourier (1768–1830).

[7] *Schrödinger's equation* for the *wavefunction* $\psi(x,t)$ of a particle of mass m in a potential $V(x)$ is: $-\frac{\hbar^2}{2m}\frac{\partial^2\psi}{\partial x^2} + V\psi = i\hbar\frac{\partial\psi}{\partial t}$, where $\hbar = 1.05457 \times 10^{-34}$ Js, which is *Planck's constant* h divided by 2π. $|\psi^2|\,dx$ is the *probability* of finding the particle in range x to $x + dx$.

Chapter 9

Radar, Chirps and Phased Arrays

9.1 Seeing with Microwaves

In 1886, Heinrich Hertz demonstrated that a spark of electricity could be used to produce electromagnetic waves. They could be reflected, refracted and diffracted, just like sound waves, but unlike sound waves, they propagate at tremendous speed, about $3 \times 10^8 \, \mathrm{ms}^{-1}$, a million times faster than sound waves in air. Electromagnetic waves can also propagate in a vacuum. A young Guglielmo Marconi, the son of Guiseppe, a wealthy Italian landowner, and Annie Jamieson, of the Irish whiskey making dynasty, had just been rejected by the navy and was experimenting with Hertz's spark-gap transmitter in his parent's attic. He was under the tutelage of his neighbour, Professor Augusto Right, one of a rather select group of experts in the new physics of *Electromagnetism*. Marconi had the idea of using *radio waves* to *communicate wirelessly*. Initial experiments indicated that these electrical waves could be detected many kilometres distant. This was achieved by his brother Alfonso confirming reception of a message 21 km away via the firing of a shotgun.[1] However, unlike the air pressure disturbance produced by the shotgun, the speed of electromagnetic waves means the possibility of near-instantaneous communication on a human scale. Given the radius of the Earth is about $R_{\oplus} = 6.371 \times 10^6 \, \mathrm{m}$, it would take radio waves just 0.134 s to make a full circle, rather better than the 80 days of Jules Verne's protagonist Phileas Fogg in 1872. Marconi pursued his idea with enormous enthusiasm over the next 40 years, and became one of the most important entrepreneurs of the early twentieth century, revolutionizing communications technology. There is a nice piece of Marconi history a few miles west of where I was raised on the Isle of Wight, a small but perfectly formed diamond shaped island a few kilometres south of the English mainland. Marconi hired the Royal Needles Hotel in Alum Bay, at the extreme western tip of the Island, for the winter season

---

[1]http://www.barrym0iow.co.uk/marconi/ (accessed 13/4/2020) and Wander, T., *Marconi on the Isle of Wight* [57].

of 1897 and constructed an experimental radio station with the goal of transmitting messages to shipping in the Solent and beyond. Distances of 64km were achieved, and Marconi's system began to gather powerful admirers, the most famous being Queen Victoria who was apparently able to communicate with the Royal Yacht from her beloved holiday residence at Osborne, which is in East Cowes in the north of the Isle of Wight. In 1903 and 1904, the German engineer Hülsmeyer [52] experimented with using radio waves to detect the presence of ships. Rather than establishing a communications link, the reception of radio waves reflected off the metal hulls of distant shipping could be used to determine their position, and via the reception of multiple transmissions, their velocity, without any special equipment being installed on the ship. It is analogous to the *sonar* (sound wave) ranging systems used by bats and aquatic mammals such as dolphins. The prey of a bat, or the walls of a cave, don't need to have special ultrasonic transducers implanted before a hunting trip commences.

The idea of radar (**ra**dio **d**etection **a**nd **r**anging), the detection of a physically remote object via the reception and processing of backscattered electromagnetic waves, was developed rapidly throughout the twentieth century, not least by Marconi and his rapidly expanding international corporation. The military utility of radar was realised early on, and systems for aircraft detection in particular (such as 'Chain Home' in the UK) played a vital role in the Second World War. One of the key innovations was the refinement of the *cavity magnetron* by John Randall and Harry Boot of the University of Birmingham in 1940. They invented a device which could produce microwaves at around 10 cm wavelength (i.e. about 3 GHz frequency) with powers of several kilowatts. The immediate application was the transformation, and miniaturization, of practical radar systems, although most people will be more familiar with the magnetron at the heart of compact domestic ovens in millions of kitchens worldwide. Domestic microwaves work on the basis of heating due to the rapid vibration of polar molecules such as water, in the presence of an electric field that oscillates at high frequency. A few billion cycles per second of these microwaves will heat up last night's leftovers to perfection.[2]

Radar is ubiquitous in the 21st century. Every commercial airport will have access to a radar feed which enables air traffic controllers to track and direct aircraft up to hundreds of kilometres away. A combination of GPS and direct radio communication may well provide a more accurate description of aircraft position, velocity, etc., but *primary radar* is still a vital backup if these systems fail, or are deliberately turned off. As the engineers at Lockheed Martin and Northrop Gruman who developed *Stealth* (i.e. low radar reflectivity) aircraft such as the F-117 Nighthawk and B-2 Spirit would surely attest, it is very difficult to hide a, *flyable*, high-speed winged

---

[2]According to the Wikipedia page on Microwave ovens (accessed 13/04/2020), a frequency of about 2.45 GHz, i.e. 12.2 cm vacuum wavelength, is typical for magnetrons in domestic use.

metal tube from microwaves. Radars work in darkness and can penetrate haze, fog, snow and rain. Attenuation due to the presence of water in the atmosphere is the basis of weather radar, and hence a key component of meteorological forecasting systems.[3] Most ships will have at least one radar on their mast, from the frisbee-sized marine radars on small yachts, all the way to the Sampson multi-function phased array radar that comprises the key sensor of the *Sea Viper* missile defence system aboard the six *Daring* class Type-45 destroyers in the UK Royal Navy. Most automobiles will now have parking sensors based upon low-power radars, and future self-driving cars will surely use radar as one of their primary sensors to aid navigation and avoid collisions. Another key feature of electromagnetic waves, is that they can propagate in the vacuum of space; *they themselves move*. No 'luminiferous aether' medium is required. This is crucial for the operation of satellites and our (only) means of communicating with spacecraft. In addition to imaging systems over the whole scope of the electromagnetic spectrum, radars are also to be found in many satellites and space probes, and have been used to accurately map the topography of not just Earth, but other rocky planets and asteroids. With the exception of a small number of man-made objects that have made physical contact with bodies within the Solar System, electromagnetic waves have been, until recently, our only means of gaining information about the wider cosmos. In recent years, gravitational wave instruments such as LIGO have added an additional mechanism. Using laser interferometry it is possibly to detect tiny ripples in gravitational fields resulting from extreme (and thankfully very distant) cosmic events such as the merger of Black Holes [1].

The applications of radar, and associated technical details, are truly enormous. In *Science by Simulation* my intention is to give the merest flavour of a few mathematical models which illustrate a few key radar ideas. For a considerably more in-depth treatment, the classic text is *Introduction to Radar Systems* [46], (the little brother to the seriously comprehensive *Radar Handbook*) by Skolnik [47]. I also recommend *Fundamentals of Radar Signal Processing* [39] by Richards, and *Introduction to Radar Target Recognition* [52] by Tait.

In the sections below we will look at three models. They cover radar propagation essentials (the *radar range equation*), antennae (*phased arrays*) and aspects of signal processing (*chirps and matched filtering*).

(1) **The radar range equation:** Given the transmitted power, frequency, pulse duration, number of pulses of a transmitter, the size of an antenna, the radar cross-section of a target, propagation losses and receiver noise characteristics,

---

[3] A *Doppler radar*, which makes use of the phase differences of reflections originating from subsequent radar pulses, can be used to remotely measure radial velocity. It is therefore a vital tool not just in aircraft detection in a *cluttered* environment, but the basis of speed-cameras and monitoring of wind velocity in hurricanes and tornados.

one can calculate how the expected signal-to-noise ratio (SNR) varies with target range R, and hence the overall performance of a ranging system.

(2) **A phased array:** Radar antennae can be used to produce a directional beam of microwaves, i.e. to focus energy on a target. However, to scan a beam by mechanically rotating an antenna can be slow, too slow to track fast-moving, or multiple targets. A *phased array* is an antenna constructed from lots of mini antennae. The microwave signal produced from each element will have a predefined phase relationship imposed, relative to all the other elements, which has the effect of steering the beam formed from the superposition of the radiation from the mini antenna array elements. The 'phase weightings' can be changed as fast as an electrical signal can be transmitted, so this means a phased array can scan a beam *much* faster than an antenna can rotate.

(3) **Chirps and matched filtering:** In order to achieve good radar range resolution, one must transmit a short pulse of microwaves. Since microwave echoes travel at the speed of light c, the difference in range is equal to $\frac{1}{2}c$ multiplied by the difference in the there-and-back time. Note the minimum of the latter is the pulse width τ, so short pulses are desirable for an additional reason: the minimum detectable range[4] is $\frac{1}{2}c\tau$. The *radar range equation* implies that these short pulses must be of very high power in order to deliver sufficient energy to the target to achieve a suitable SNR, and hence enable detection at ranges of possibly hundreds of kilometres. Unfortunately a high power transmission places significant stress upon the components of a radar system, and increases its size and cost. An alternative is to deliver the same energy via a longer pulse that *varies in a characteristic way with frequency*. This frequency coding gets around the worsening of range resolution as pulse width increases, although augments the minimum range problem. A linear 'chirp' (i.e. a pulse whose frequency increases or decreases at a constant rate) is possibly the simplest example. A signal processing technique called *cross-correlation* will compare the transmitted pulse with what is received, and synthesise mathematically what is essentially the same as reflections from much shorter pulses of equivalent energy.

9.2 The Radar Equation

The power per unit area at range R from an isotropic source of radiation that transmits power P_t is $\frac{P_t}{4\pi R^2}$, i.e. the total power transmitted divided by the surface area of a sphere. Assume a target (e.g. an aeroplane, a ship, a bird, a rain cloud, or indeed a planet) of cross-sectional area σ receives the radiation, and in the direction of the target, the radiation source (which is actually an antenna, with some degree

---

[4]This assumes a *monostatic* system, i.e. the same radar antenna is used to (alternately) transmit, then receive. During the transmit time τ the radar cannot receive any reflections.

of directivity), confers *gain G*. Also assume there is some form of loss factor L_t due to inefficiencies in the generation of radiation, and a propagation factor F which relates to the effect of the environment on the radiation between the source and the target. This could be a loss (e.g. absorption of radiation by atmospheric gases[5]) or possibly a gain, if there is constructive interference between direct and indirect ray paths to a target (e.g. from ground reflection). Incorporating all these factors, the radiative power P_R received by the target is:

$$P_R = \frac{\sigma}{4\pi R^2} G P_t \frac{F}{L_t} \qquad (9.1)$$

The maximum gain of an antenna is given by the equation [39, p. 13]:

$$G = \frac{4\pi\eta A}{\lambda^2} \qquad (9.2)$$

where λ is the wavelength of the radiation, A is the cross-sectional area of the antenna and η is the 'antenna efficiency', a geometrical factor particular to the design of the antenna. Using $c = f\lambda$, we can express this in terms of the frequency f of radiation and the speed of radiation c. (For electromagnetic waves, this will be the speed of light, which in a vacuum is $2.998 \times 10^8 \, \mathrm{ms^{-1}}$).

$$\therefore \ G = \frac{4\pi\eta A f^2}{c^2} \qquad (9.3)$$

Note in terms of the angular widths of the primary radar beam $\Delta\phi$ in azimuth and $\Delta\varepsilon$ in elevation, the antenna gain is: [46, p. 77]

$$G \approx \frac{26,000}{(\Delta\phi/\deg)(\Delta\varepsilon/\deg)} \qquad (9.4)$$

Now assume the target radiates isotropically[6] back towards the radar, which can be used to alternately transmit and then listen for reflections for a short period.[7] The reflected power received by the antenna of effective receiving area ηA is, again taking into account propagation and receiver losses:

$$P_r = \frac{\eta A}{4\pi R^2} P_R \frac{F}{L_r} \qquad (9.5)$$

---

[5]The scattering of electromagnetic waves by water droplets is actually a very useful phenomena when using radar to detect weather such as rain.

[6]*Not* radiating isotropically, especially back in the direction of the radar, is one of the basic principles of stealth technologies, used extensively in modern military aircraft, ships and land vehicles. A faceted surface will often suffice, which partially explains why the Lockheed F-117 Nighthawk military aircraft has the distinctively angular airframe. Unless the radar happens to be aligned with a surface normal of the aircraft, the law of reflection states that radiation will tend to be reflected away from the source, thus reducing its effective radar cross-section (RCS) σ.

[7]This is the basic principle of a *monostatic, primary* radar, i.e. a transmitter and receiver located in a single system, and using the same antenna and wave-guiding apparatus. A bistatic or *multi-static* system will have physically separate transmit and receive antennae, and a *secondary* system will listen for broadcasts only, rather than specifically tune in to reflections from its own transmissions.

Hence, in terms of transmission, the received power diminishes as $1/R^4$ due to the inverse-square loss in both radar to target, *and* target to radar paths.

$$P_r = \frac{\eta A}{4\pi R^2} \left(\frac{\sigma}{4\pi R^2} GP_t \frac{F}{L_t} \right) \frac{F}{L_r}$$

$$P_r = \frac{\sigma \eta AGP_t F^2}{16\pi^2 L_t L_r R^4} \tag{9.6}$$

A radar may transmit n pulses and combine the reflections from these to give further signal gain. Additionally, a pulse of duration τ and bandwidth (i.e. frequency extent) B can result in an additional gain of $B\tau$ if an ideal *matched filter* (see the last section of this chapter) is used to correlate the received signals with what was transmitted. The received power from n of these pulses, combined with 'efficiency factor' $E_i(n)$ is:

$$P_r = \frac{\sigma \eta AGP_t F^2}{16\pi^2 L_t L_r R^4} nE_i(n)B\tau \tag{9.7}$$

Now the power of noise[8] in a radar receiver is given by [46, p. 33]:

$$P_{\text{noise}} = k_B T_0 N_f B \tag{9.8}$$

where B is the spectral bandwidth of waves, T_0 is the ambient temperature in K and $k_B = 1.38 \times 10^{-23} \text{JK}^{-1}$ is Boltzmann's constant. N_f is the *noise figure* of a receiver, and most receivers typically have a $T_0 N_f$ value of about $500\,\text{K}$.

The SNR S/N of received power relative to receiver noise is effectively what is used to define a detection in the radar system:

$$S/N = \frac{P_r}{P_{\text{noise}}} \tag{9.9}$$

Hence, using $G = \frac{4\pi A\eta f^2}{c^2}$:

$$S/N = \underbrace{\eta^2 A^2 \tau f^2 P_t}_{\text{antenna properties}} \times \underbrace{nE_i(n)}_{\text{pulses}} \times \underbrace{\frac{F^2}{L_t L_r T_0 N_f}}_{\text{loss and noise}} \times \underbrace{\frac{\sigma}{R^4}}_{\text{target and range}} \times \underbrace{\frac{1}{4\pi c^2 k_B}}_{\text{constants}} \tag{9.10}$$

This equation for S/N is known as the *radar range equation*, as it relates many of the fundamental high-level parameters of a radar system. I've grouped this possibly unwieldy product into (i) antenna properties, (ii) pulses, (iii) loss and noise terms, (iv) target and range terms, (v) physical constants. A plot of $10\log_{10}(S/N)$ vs R is illustrated in Fig. 9.1, with different coloured lines corresponding to different transmitter powers.

---

[8]Crudely speaking, if $k_B T_0$ is the mean kinetic energy of molecules, they will vibrate B times a second as a consequence of a pulse of bandwidth (i.e. frequency extent) B. Hence the power of thermal noise is going to be proportional to $k_B T_0 B$.

Note the radar antenna area A can be determined from beamwidth using the gain expression: $G = \frac{4\pi A \eta f^2}{c^2} \approx \frac{26,000}{(\Delta\phi/\deg)(\Delta\varepsilon/\deg)}$.

$$\therefore A = \frac{26,000}{(\Delta\phi/\deg)(\Delta\varepsilon/\deg)} \frac{c^2}{4\pi f^2 \eta} \tag{9.11}$$

(a)

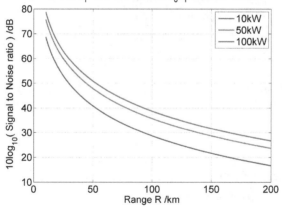

(b)

Fig. 9.1 (a) The *radar range equation* for the ratio of received power of microwaves reflected off a target such as an aircraft to the power of receiver noise is: $S/N = \eta^2 A^2 \tau f^2 P_t \times n E_i(n) \times \frac{F^2}{L_t L_r T_0 N_f} \times \frac{\sigma}{R^4} \times \frac{1}{4\pi c^2 k_B}$. The inverse fourth-power with range R is due to the microwaves being transmitted by an antenna, reflected off the aircraft of radar cross-section σ, and then the re-radiated microwaves being detected by the same antenna. This is a common, *monostatic* situation, and requires the radar to be set in alternate transmit and receive modes on a regular time schedule. Although both the radar antenna and the aircraft will be directional in some sense (i.e. not radiating isotropically in all directions), an inverse-law with range is assumed for each direction, hence an inverse-fourth power law overall. Graph (b) is a plot of signal to noise ratio in decibels (dB). $10\log_{10}(S/N)$ is plotted against range R for various transmitter powers in kW. f_c is the transmitter frequency, A is the antenna area, η is the 'antenna radiating efficiency', $T_0 N_f$ is the 'noise temperature' of the receiver, L is the total loss during propagation and reception. It is assumed n pulses of length τ are combined with efficiency E_i to form a detection.

9.3 The Phased Array Antenna

9.3.1 *A linear array*

Consider a line of N isotropic radiating elements, with radiation wavelength λ, separated by fixed spacing s. The x coordinate of array element $n = 1 \ldots N$ is given by:

$$x_n = -\tfrac{1}{2}Ns + (n-1)s \tag{9.12}$$

However, for various practical reasons it might not be possible to arrange array elements in a perfect line, so in general we shall consider a set $\{x_n, y_n\}$, as in Fig. 9.2. The deviations from above could be small, random errors, or perhaps a deliberately curved system such as a circular array.

Consider an azimuth angle θ measured from the normal to the line-array. The far-field wavefield $\psi(R, \theta, t)$ at range R from the centre of the array is given by

$$\psi(R, \theta, t) = \sum_{n=1}^{N} w_n e^{2\pi i \frac{\Delta R_n}{\lambda}} e^{-i\omega t} \tag{9.13}$$

Note we are using *de-Moivre's theorem* to represent the propagation phase shift in terms of a complex exponential factor $e^{2\pi i \frac{\Delta R_n}{\lambda}}$, as discussed at the end of the previous chapter. The time variation of the wavefield is $e^{-i\omega t}$, i.e. also complex-exponential, to represent a periodic variation of frequency $f = \frac{\omega}{2\pi} = \frac{c}{\lambda}$. The distance ΔR_n from array element x_n to (R, θ) can be determined via the Cosine Rule (see Fig. 9.2(b)). This yields:

$$\Delta R_n = \sqrt{R^2 + x_n^2 + y_n^2 - 2R\sqrt{x_n^2 + y_n^2}\cos\alpha_n} \tag{9.14}$$

where

$$\alpha_n = \frac{\pi}{2} - \theta - \tan^{-1}\left(\frac{y_n}{x_n}\right) \tag{9.15}$$

ΔR_n can be simplified by considering a binomial expansion of $\frac{\Delta R_n}{\lambda}$ when in the 'far field', i.e. when $R \gg \lambda$, and also $R \gg Ns$:

$$\Delta R_n = R\sqrt{1 + \frac{x_n^2 + y_n^2}{R^2} - \frac{2\sqrt{x_n^2 + y_n^2}}{R}\cos\alpha_n}$$

$$\Delta R_n \approx R\left(1 - \frac{2\sqrt{x_n^2 + y_n^2}}{R}\cos\alpha_n\right)^{\frac{1}{2}}$$

$$\Delta R_n \approx R - \sqrt{x_n^2 + y_n^2}\cos\alpha_n \tag{9.16}$$

i.e. we can neglect the $\frac{x_n^2 + y_n^2}{R^2}$ term compared to the others. The expression above is used to compute the set of (complex) weights w_n which are applied to each array element, that enables a beam to be positioned in azimuth. Since this can be done at the speed of electrical signals in a radar, this means a phased array antenna can be steered (within certain limits) considerably faster than a system which sweeps a beam via mechanical rotation only. This is a good reason why phased arrays are often found in systems which are used for defence against very fast-moving objects

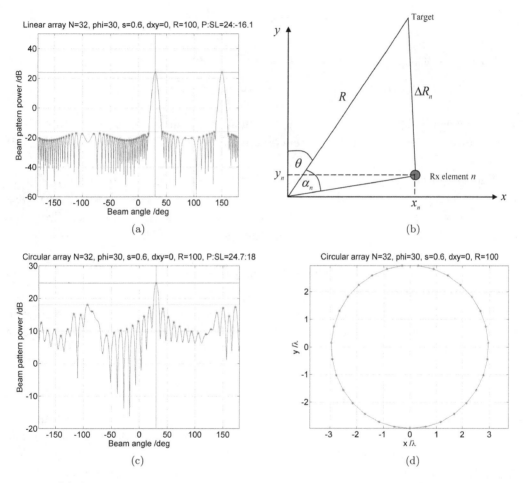

Fig. 9.2 (a) A model of the beam pattern of a linear phased array antenna of $N = 32$ radiating elements separated by $s = 0.6\lambda$ spacing. A weighting is applied to the array elements, depicted in (b), to result in a main beam scan of $\phi = 30°$, and a *sidelobe* level of about 40 dB below the main beam peak. There is also a repeated *grating lobe* at $\varphi = 180° - 30° = 150°$, but we can neglect this since the scan angle is beyond 90° from 'boresight' and the antenna may be designed to absorb (or reflect) radiation that doesn't propagate forward. (c) Similar phased array to (a), but now a circular geometry, as illustrated in (d). The circular symmetry means the main beam should remain similarly sharp regardless of scan angle, but unfortunately the sidelobe level is higher than the linear array.

such as missiles. To steer by azimuth angle ϕ, the (complex) weighting is:

$$w_n = a_n e^{i\Delta\Omega_n} \tag{9.17}$$

The phase weighting part of $w_n = a_n e^{i\Delta\Omega_n}$ is:

$$\Delta\Omega_n = \frac{2\pi}{\lambda}\sqrt{x_n^2 + y_n^2}\cos\beta_n \tag{9.18}$$

$$\beta_n = \frac{\pi}{2} - \phi - \tan^{-1}\left(\frac{y_n}{x_n}\right) \tag{9.19}$$

When $\theta = \phi$, $\alpha_n = \beta_n$, the product $w_n e^{2\pi i\frac{\Delta R_n}{\lambda}}$ is

$$e^{2\pi i\frac{\Delta R_n}{\lambda}} e^{i\Delta\Omega_n} = e^{\frac{2\pi i}{\lambda}\left(R - \sqrt{x_n^2+y_n^2}\cos\alpha_n + \sqrt{x_n^2+y_n^2}\cos\beta_n\right)} = e^{\frac{2\pi i}{\lambda}R} \tag{9.20}$$

which means all elements contribute the same, and are in-phase, which is essentially the criteria for constructive interference. Therefore we expect a beam maximum at $\theta = \phi$.

The *amplitude weighting* a_n, used to minimise *sidelobes* due to the finite number N of array elements, will typically take an *amplitude taper form*. This means weighting contributions from array elements nearer the centre, and less so from the outer edges. A *Hamming* taper is one of the simplest formulae:

$$a_n = 0.54 - 0.46\cos\left(\frac{2\pi(n-1)}{N}\right) \tag{9.21}$$

The beam pattern power (in decibels) is $10\log_{10}|\psi|^2$, and this is plotted vs θ in Fig. 9.2. Linear and circular phased arrays are compared, and a scan angle of $\phi = 30°$ is set.

9.3.2 *A 2D phased array*

A similar argument can be used to determine the weightings for a 2D rectangular array. Consider an $N \times M$ array of isotropic radiating elements of spacing s. The position vector of a target at range R, azimuth ϕ and elevation ε from the centre of the array is, in x, y, z coordinates (with the array defined in the x, y plane)

$$\mathbf{R} = R\left(\cos\varepsilon\sin\phi, \sin\varepsilon, \cos\varepsilon\cos\phi\right) \tag{9.22}$$

The wavefield $\psi(R, \phi, \varepsilon, t)$ at range R from the centre of the array is again given by:

$$\psi(R, \phi, \varepsilon, t) = \sum_{n=1}^{N}\sum_{m=1}^{M} w_{n,m} e^{2\pi i\frac{\Delta R_{n,m}}{\lambda}} e^{-i\omega t} \tag{9.23}$$

where $\Delta R_{n,m} = |\mathbf{a}_{n,m} - \mathbf{R}|$ and $\mathbf{a}_{n,m} = (x_n, y_m, 0)$.

Hence, making good use of $\sin^2 \varepsilon + \cos^2 \varepsilon = 1$:

$$\Delta R_{n,m} = \sqrt{\left(R \cos \varepsilon \sin \phi - x_n\right)^2 + \left(R \sin \varepsilon - y_m\right)^2 + R^2 \cos^2 \varepsilon \cos^2 \phi}$$

$$\Delta R_{n,m} = \sqrt{\begin{array}{l} R^2 \cos^2 \varepsilon \sin^2 \phi + -2x_n R \cos \varepsilon \sin \phi + x_n^2 + R^2 \sin^2 \varepsilon - 2y_m R \sin \varepsilon + y_m^2 \\ + R^2 \cos^2 \varepsilon \cos^2 \phi \end{array}}$$

$$\Delta R_{n,m} = \sqrt{R^2 - 2x_n R \cos \varepsilon \sin \phi - 2y_m R \sin \varepsilon + x_n^2 + y_m^2}$$

$$\Delta R_{n,m} = R\sqrt{1 - \frac{2x_n}{R} \cos \varepsilon \sin \phi - \frac{2y_m}{R} \sin \varepsilon + \frac{x_n^2 + y_m^2}{R^2}}$$

$$\Delta R_{n,m} \approx R \left(1 - \frac{2}{R} \left(x_n \cos \varepsilon \sin \phi - y_m \sin \varepsilon\right)\right)^{\frac{1}{2}}$$

$$\Delta R_{n,m} \approx R - x_n \cos \varepsilon \sin \phi - y_m \sin \varepsilon \tag{9.24}$$

which means to steer a beam to azimuth Φ and elevation Ξ:

$$w_{n,m} = a_n b_m e^{i \Delta \Omega_{n,m}} \tag{9.25}$$

$$\Delta \Omega_{n,m} = \frac{2\pi}{\lambda} \left(x_n \cos \Xi \sin \Phi - y_m \sin \Xi\right) \tag{9.26}$$

$$a_n = 0.54 - 0.46 \cos \left(\frac{2\pi(n-1)}{N}\right) \tag{9.27}$$

$$b_m = 0.54 - 0.46 \cos \left(\frac{2\pi(m-1)}{M}\right) \tag{9.28}$$

The beam pattern power is $10 \log_{10} |\psi|^2$, and this is plotted vs ϕ and ε in Fig. 9.3 for a 16×16 array. Scan angles of $\phi = 15°$, $\varepsilon = 15°$ are compared to a 'boresight' $\phi = 0°$, $\varepsilon = 0°$.

9.4 Chirps and Matched Filtering

9.4.1 *Linear chirp*

A linear chirp pulse waveform $\psi(t)$ is a rectangular envelope bounded signal with a frequency that increases in direct proportion to the time since the leading edge of the pulse (see Fig. 9.4). Let the chirp begin at time t_0 and have duration τ. Let the frequency range (bandwidth) B extend symmetrically about a mean (i.e. the 'carrier frequency') $f_c = \frac{c}{\lambda}$. The chirp instantaneous frequency $f(t)$ can therefore be expressed:

$$f(t) = \begin{cases} a + bt & t_0 < t < t_0 + \tau \\ \text{undefined} & t < t_0, \quad t > t_0 + \tau \end{cases} \tag{9.29}$$

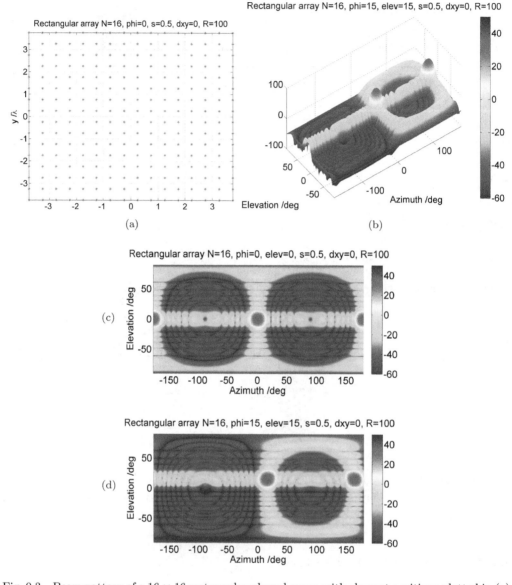

Fig. 9.3 Beam pattern of a 16 × 16 rectangular phased array, with element positions plotted in (a). A suitable phase weighting can enable the main beam to be steered in both azimuth and elevation, and therefore a radar based upon this type of antenna can detect targets in three-dimensions. In (b) and (c), the beam power in dB is plotted vs azimuth ϕ and elevation ε, and phase weightings are such that the main beam points at $\phi = 0°, \varepsilon = 0°$. Note the presence of a large grating lobe separated by $\Delta\phi = 180°$ –the scan angle. In reality the array will probably be shielded so it can't receive beyond scan angles of about $\pm 90°$, so a problem of fore-and rear confusion is unlikely to manifest. Plot (d) illustrates the situation where an array phase weighting can steer the beam to $\phi = 15°, \varepsilon = 15°$.

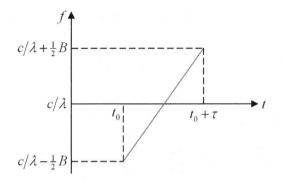

Fig. 9.4 A linear chirp is a frequency coded signal that increases in frequency by bandwidth B over the duration τ of the waveform. f_c is the carrier frequency of the signal.

Now:

$$f(t_0) = a + bt_0 = \frac{c}{\lambda} - \tfrac{1}{2}B \tag{9.30}$$

$$\frac{df}{dt} = b = \frac{B}{\tau} \tag{9.31}$$

Hence:

$$a = \frac{c}{\lambda} - B\left(\tfrac{1}{2} + \frac{t_0}{\tau}\right) \tag{9.32}$$

$$b = \frac{B}{\tau} \tag{9.33}$$

which gives:

$$f(t) = \begin{cases} \dfrac{c}{\lambda} + B\left(\dfrac{t - t_0}{\tau} - \tfrac{1}{2}\right) & t_0 < t < t_0 + \tau \\ \text{undefined} & t < t_0, \quad t > t_0 + \tau \end{cases} \tag{9.34}$$

The chirp signal amplitude $\psi_{\text{real}}(t) = A\cos(\phi(t)) = A\operatorname{Re}\left\{e^{i\phi}\right\}$, where ϕ is the phase of the waveform and A is defined to be a real constant. Let us define $\psi(t) = e^{i\phi}$ to be the (unit amplitude) complex part of this mathematical representation of the signal $\psi_{\text{real}}(t)$. The complex part $\psi(t)$ shall be referred to as the 'chirp signal' from now on.

$$\psi(t) = e^{i\phi(t)} \tag{9.35}$$

$$\psi_{\text{real}}(t) = A\operatorname{Re}\left\{\psi(t)\right\} \tag{9.36}$$

The phase ϕ is related to instantaneous frequency (in the continuous region of the pulse envelope $t_0 < t < t_0 + \tau$) by:

$$f(t) = \frac{1}{2\pi} \frac{d\phi}{dt} \tag{9.37}$$

Hence:

$$\phi(t) = 2\pi \int f(t)dt \tag{9.38}$$

Conventionally (9.38) is an indefinite integral with constant of integration γ set such that the phase at the centre of the pulse is zero.

$$\phi\left(t_0 + \tfrac{1}{2}\tau\right) = 0 \tag{9.39}$$

Using (9.34) and the above condition, the chirp phase can be evaluated analytically:

$$\phi(t) = 2\pi \int \left(\frac{c}{\lambda} + B\left(\frac{t - t_0}{\tau} - \tfrac{1}{2} \right) \right) dt$$

$$= 2\pi \left[\frac{c}{\lambda} t + Bt \left(\frac{\tfrac{1}{2}t - t_0}{\tau} - \tfrac{1}{2} \right) \right] + \gamma$$

$$= 2\pi \left[\tfrac{1}{2} \frac{B}{\tau} t^2 - \left(\frac{Bt_0}{\tau} + \tfrac{1}{2}B - \frac{c}{\lambda} \right) t \right] + \gamma \tag{9.40}$$

Since $\phi\left(t_0 + \tfrac{1}{2}\tau\right) = 0$:

$$\gamma = -2\pi \left[\tfrac{1}{2}\frac{B}{\tau} \left(t_0 + \tfrac{1}{2}\tau\right)^2 - \left(\frac{Bt_0}{\tau} + \tfrac{1}{2}B - \frac{c}{\lambda} \right) \left(t_0 + \tfrac{1}{2}\tau\right) \right] \tag{9.41}$$

Samples of the chirp (following analogue to digital conversion) occur after a *mix down* by some frequency f_{mix}. In this case:

$$f \longrightarrow f - 2\pi f_{\text{mix}} \tag{9.42}$$

If $f_{\text{mix}} = \frac{c}{\lambda}$ then the signal is said to be converted down to 'baseband'.

In summary, the linear chirp is described by the following system of equations:

$$\psi_{\text{real}}(t) = A\,\text{Re}\left\{ \psi(t) \right\}$$

$$\psi(t) = \begin{cases} e^{i\phi(t)} & t_0 < t < t_0 + \tau \\ 0 & t < t_0, \quad t > t_0 + \tau \end{cases}$$

$$\phi(t) = \alpha t^2 + \beta t + \gamma$$

$$\alpha = \frac{\pi B}{\tau} \tag{9.43}$$

$$\beta = -2\pi \left(\frac{Bt_0}{\tau} + \tfrac{1}{2}B - \frac{c}{\lambda} \right)$$

$$\gamma = -2\pi \left[\tfrac{1}{2}\frac{B}{\tau} \left(t_0 + \tfrac{1}{2}\tau\right)^2 - \left(\frac{Bt_0}{\tau} + \tfrac{1}{2}B - \frac{c}{\lambda} \right) \left(t_0 + \tfrac{1}{2}\tau\right) \right]$$

9.4.2 *Matched filter*

The *cross-correlation* between receiver signal $r(t)$ and transmitted signal $s(t)$ is the *convolution* of $r(t)$ and the time-reversed, complex conjugate of $s(t)$.

$$\chi(t) = r(t) * s^*(-t) = \int_{-\infty}^{\infty} r(t - t')s^*(-t')dt' \tag{9.44}$$

In order to implement a cross-correlation operation as a *matched filter* in a radar signal processing chain, this integral needs to be evaluated efficiently in a computational sense. This can be achieved using the idea that Fourier Transforms (and inverse Fourier Transforms) can be implemented in computer software (or indeed bespoke hardware) very efficiently as a Fast Fourier Transform (FFT) (*Numerical Recipes* [34, pp. 501+]). The Fourier transform $\hat{S}(\omega)$ of a time signal $s(t)$ is:

$$\hat{S}(\omega) = \Im[s(t)] = \frac{1}{\sqrt{2\pi}} \int_{-\infty}^{\infty} s(t)e^{-2\pi i\omega t}dt \tag{9.45}$$

and the inverse of this process is:

$$s(t) = \Im^{-1}[\hat{S}(\omega)] = \frac{1}{\sqrt{2\pi}} \int_{-\infty}^{\infty} \hat{S}(\omega)e^{2\pi i\omega t}d\omega \tag{9.46}$$

The *convolution theorem* of Fourier transforms states that:

$$\Im[a * b] = \Im[a] \times \Im[b] \tag{9.47}$$

which means our matched filter can be efficiently implemented by performing the inverse FFT of the product of the FFTs of (i) the time-reversed, complex conjugate of the transmitted signal and (ii) the received signal. Both of these can be regarded as a vector of complex numbers, which correspond to samples of the respective signals at some fixed rate.

$$\chi(t) = r(t) * s^*(-t) = \Im^{-1}\left[\Im[r(t)] \times \Im[s^*(-t)]\right] \tag{9.48}$$

$\Im[s^*(-t)]$ could be computed once, and stored for regular future retrieval in the memory of the matched filter signal processor. Code for a MATLAB implementation of a matched filter is given below. Note % prefixes mean commentary rather than code, and `hamming(N)` is another MATLAB function which generates a set of N weights $w_m = 0.54 - 0.46\cos\left(\frac{2\pi(m-1)}{N}\right)$.

```
%xcor
% Function which computes cross-correlation with transmitted signal tx
% and received signal rx.
function [xc,t] = xcor( tx,rx,dt )
% Time reverse and complex conjugate transmitted signal and pad with zeros.
% Also apply a window to reduce effects of the finite width
% DFT as compared to a full Fourier Transform.
N = round( length(rx)/2 ); w = hamming( length(tx) );
x = [ zeros(1,N),fliplr( conj( tx.*w ) ),zeros(1,N) ];
```

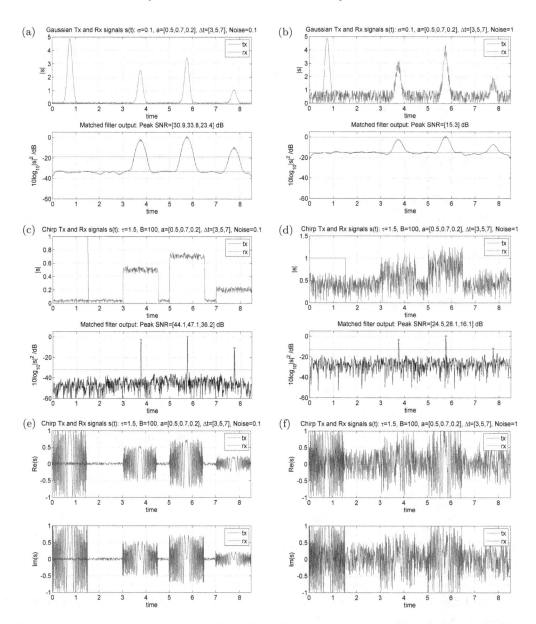

Fig. 9.5 Demonstration of the use of matched filtering or 'pulse-compression' to result in an improved signal-to-noise ratio. A matched filter cross-correlates what is received to what was transmitted. In all examples, a radar pulse is reflected off three targets which have slightly different reflection amplitudes a, and are at different ranges. The left graphs have a relative noise amplitude of 0.1, whereas the right graphs have a noise amplitude of 1. Some radars transmit short pulses, as in (a) and (b). However, in order for targets to be localised in range, the pulse widths needs to be very short, which presents a problem transmitting enough energy to obtain a suitable SNR. An alternative is to transmit a longer, frequency coded pulse such as a 'chirp', as plotted in (e) and (f). The output of the matched filter, (c) and (d), has the same effect of transmitting a short pulse with the same energy as the longer chirp.

```
% Pad x or rx with zeros
if length(x)>length(rx)
 y = [rx,zeros( 1,length(x)-length(rx) )];
elseif length(rx)>length(x)
 x = [x,zeros( 1,length(rx)-length(x) )];
end
% Perform convolution of x and y using FFT functions
xc=ifftshift( ifft( fft(x).*fft(y) ) );
% Generate corresponding time vector for matched filter output
t = 0:dt:( length(xc)-1 )*dt;
```

The matched filtering concept is illustrated in Fig. 9.5. A matched filter cross-correlates what is received to what is transmitted. In all examples, a radar pulse is reflected off three targets which have slightly different reflection amplitudes a, and are at different ranges. The left graphs have a relative noise amplitude of 0.1, whereas the right graphs have a noise amplitude of 1.0. Figures 9.5(a) and 9.5(b) correspond to short 'Gaussian' pulses. Longer, frequency coded linear chirp pulses are plotted in Figs. 9.5(e) and 9.5(f). The output of the matched filter, Figs. 9.5(c) and 9.5(d), has the same effect of transmitting a much shorter pulse with the same energy as the longer chirp. In other words, we can obtain a good range resolution from a much longer pulse if we using chirp frequency coding.

One can show (e.g. [39, p. 196]) that the range resolution using a matched filter and a linear chirp of bandwidth B is:

$$\Delta R = \tfrac{1}{2}\frac{c}{B} \tag{9.49}$$

whereas a short pulse of duration τ will have range resolution $\Delta R = \frac{1}{2}c\tau$. To achieve a range resolution of $\Delta R = 1.0$ m means a chirp of bandwidth $B = 149.9$ MHz, which is achievable[9] given typical carrier frequencies of several gigahertz. A typical chirp pulse length might be of the order of $20\,\mu$s. Compare this to a short, non-frequency coded pulse. To achieve $\Delta R = 1.0$ m means $\tau = 0.0067\,\mu$s.

In Fig. 9.5, the chirp has a bandwidth of $\tau = 1.5$ time-units and a bandwidth of $B = 100$ inverse-time-units. This allows for some generality, i.e. these could be μs and MHz, or ms or kHz, or s and Hz. The pulse width of the output of the matched filter is effectively $\frac{1}{B} = 0.01$ time-units. Figures 9.5(c) and 9.5(d) demonstrate that the matched filter process is pretty tolerant to noise, impressively given the characteristic waveform shape of the linear chirp in Fig. 9.5(e) is quite hard to discern (visually) from the noisy version in Fig. 9.5(f).

---

[9]Although imposes a challenge for a phased array radar. A wide-bandwidth chirp may scan the beam off target during the transmission of the pulse. The *synthesis* of wide-bandwidth waveforms that can offer high range resolution, deliverable by a phased array radar, was a large part of my PhD thesis [14].

Chapter 10

Navigating the Sphere

To a very good approximation, the Earth is spherical.[1] This assumption regarding the geometry of the Earth can be used to calculate hard-to-measure parameters such as (i) the radius $R_\oplus$ of the Earth and (ii) the distance d to the horizon when at altitude h. The calculation of $R_\oplus$, combined with a measurement of the mean density ρ of rocks, enables one to estimate of the mass $M_\oplus$ of the Earth via the equation:

$$M_\oplus = \tfrac{4}{3}\pi R_\oplus^3 \rho \qquad (10.1)$$

Problems associated with the large-scale geometry of the Earth are the basis of the modern scientific discipline of *geodesy*, and none of these are particularly difficult to solve once you have satellites in orbit with radar, laser rangefinders and sensitive gravitometers. The Earth's radius is $R_\oplus = 6.371 \times 10^6$ m and average rock density is $\rho = 5.513$ kg, which means $M_\oplus = 5.972 \times 10^{24}$ kg.

However, how might you calculate the radius of the Earth, *without* such modern technology? And indeed how did the Ancients demonstrate that the Earth was approximately spherical?

10.1 Eratosthenes of Cyrene Calculates the Radius of the Earth

Eratosthenes of Cyrene (276BC–194BC), Greek polymath and former chief librarian at the Library of Alexandria, invented the discipline of *Geography*, and is credited with the first relatively accurate measurement of the radius of the Earth. The curvature of the Earth is only just perceivable from the summit of the highest mountains. A theory of a spherical Earth is therefore rather a bold leap of the imagination for a proto-scientist of the ancient world, bereft of a deity-eyed view from satellites and space stations. It is likely such an idea may have been suggested to explain the

---

[1]The polar radius differs from the equatorial radius by a factor of $\approx 1 - \frac{1}{228.257}$.

horizon problem, i.e. why the masts of ships appear to vanish when they travel far enough from shore. Eratosthenes noticed rays of summer-solstice noon sunlight cast no shadow of a well at Syene (Aswan), whereas simultaneously[2] in Alexandria, a vertical pole would cast a shadow with an angle of about 7.8° from the vertical.[3] Given the theory of a spherical Earth, and that sunlight reaches Earth as parallel rays, an explanation of this effect is that the Syene rays point towards the centre of the Earth (i.e. straight down the well) whereas the Alexandrian rays deviate from the vertical by 7.8°. If the distance of the Earth-surface arc separating Alexandria and Syene could be accurately measured, then geometry will enable one to compute the radius of the Earth, as indicated in Fig. 10.1(a).

Eratosthenes' calculation is repeated below using modern units. The 6.4% accuracy is rather impressive given the difficulty of measuring the distance between Alexandria and Syene. The journey criss-crosses the Nile and therefore would encounter many obstacles in the form of topography, civilisation, bandits and crocodiles! The circular arc on the surface of a spherical Earth which corresponds to the distance between Alexandria and Syene was measured at 5,000 *stadia*. 1 stadia is ≈ 185 m. The arc is $R\theta$ where R is the radius of the Earth and θ is the angle between Alexandra and Syene as measured from the centre of the Earth. It is *also* the angle (in radians) of the shadow at Alexandria. Hence:

$$R \approx \frac{5{,}000 \times 185\,\text{m}}{7.8 \times \frac{\pi}{180}} = 6.78 \times 10^6\,\text{m} \qquad (10.2)$$

The modern measure of Earth radius is $R_\oplus = 6.371 \times 10^6$ m, which means Earatoshenes was out by:

$$\frac{R - R_\oplus}{R_\oplus} = \frac{6.78 - 6.371}{6.371} \approx 6.4\% \qquad (10.3)$$

10.2 Al-Biruni of Khwarezm Improves Upon the Method of Eratosthenes

The legendary Islamic scholar Al-Biruni (973AD–1050AD) of Khwarezm (modern day Uzbekistan and Turkmenistan) improved upon Eratosthenes' method, as described in his *Codex Masudicus* (1037) [22] using a much shorter 'baseline', *three* angular measurements of elevation, and the use of *trigonometry*. Firstly he measured the height of a hilltop Fort Nandana (near Pind Dadan Khan in Pakistan) above flat plains, which one assumes are at sea level or close to it. The same method

---

[2]Clearly this must have been based upon observations made over several years!.

[3]I confess a bit of reverse engineering in this calculation. Eratosthenes may have assumed Syene was due South of Alexandria, which is not quite correct. If one assumes a minor arc of a Great Circle between Alexandria (31.2°N, 29.2°E) and Aswan (24.1°N, 32.9°E), then the angle between these points and the centre of the Earth is: $\theta = 7.82°$. The methodology to calculate this angle is described in Section 10.4.3.

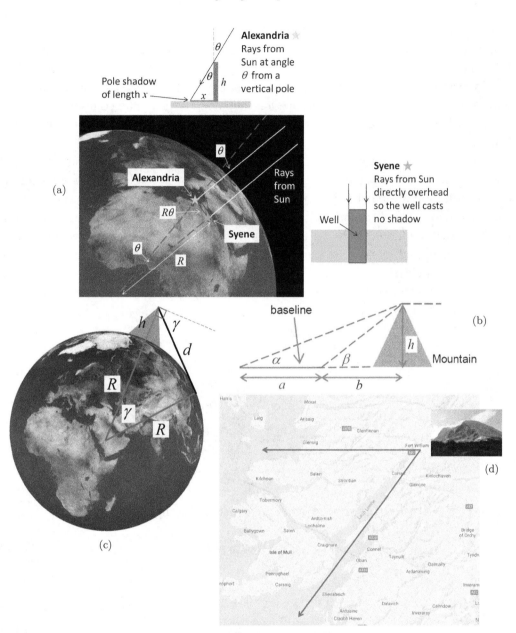

Fig. 10.1   (a) Geometry associated with Eratosthenes' calculation of the radius of the Earth, by comparing the angle of a shadow at Alexandria to a shadowless well at Syene. Note $\theta = \tan^{-1}(x/h)$. (b) Geometry of a trigonometric method to determine the height h of a mountain using a baseline a and mountain-peak elevation angles α and β taken from either ends of the baseline. (c) Al-Biruni's method can work out the Earth radius more accurately than that of Eratosthenes, as long as you can accurately measure the angle of dip γ of the horizon from a mountain peak, and know the height h of the mountain. On a clear day, it might be possible to evaluate Al-Biruni's formula from the summit of Ben Nevis (1,345m), given one may be able to see the sea from at least two directions. (d) West Coast of Scotland map image from Google Maps. Earth texture from Wikipedia: https://commons.wikimedia.org/wiki/File:Earthmap1000x500.jpg.

was, in essence, employed in the *Great Trigonometric Survey of India* (1802–1871) culminating in measuring the height of the Himalaya, with Everest[4] (8,848 m) being the highest peak. From the battlements of Nandana, Al-Biruni used an *Astrolabe* (an intricate inclinometer) to measure the angle of the horizon from the vertical, with a mass on a string as a reference for the latter. This angle, combined with the height of Fort Nandana above sea level, can be used to determine the radius of the Earth.

Let h be a mountain height, and a be the length of an accurately measured baseline which connects two points at which the elevation angles α, β of the summit of the mountain are recorded, as in Fig. 10.1(b). Furthermore, let b be the distance from the (closer) β elevation end of the baseline, and an extension of this line to a point directly underneath the mountain summit. Simple trigonometry would enable h to be calculated merely from one of the elevation angles and either $a + b$, or b. However, it is clearly impractical to drill a hole to the centre of the mountain and hence measure b directly. Hence the need for the *pair* of elevation measurements. From the definition of tan:

$$h = (a + b) \tan \alpha \tag{10.4}$$

$$h = b \tan \beta \tag{10.5}$$

$$\therefore h = a \tan \alpha + \frac{h}{\tan \beta} \tan \alpha$$

$$\therefore h \tan \beta - h \tan \alpha = a \tan \alpha \tan \beta$$

$$\therefore h = \frac{a \tan \alpha \tan \beta}{\tan \beta - \tan \alpha} \tag{10.6}$$

Referring to Al-Biruni's 'horizon triangle' construction in Fig. 10.1(c):

$$(R + h) \cos \gamma = R \tag{10.7}$$

$$\therefore h \cos \gamma = R(1 - \cos \gamma)$$

$$\therefore R = \frac{h \cos \gamma}{1 - \cos \gamma} \tag{10.8}$$

Let us illustrate Al-Biruni's method with quantities that could be readily measured in the West Highlands of Scotland, a delightful part of the world and the

---

[4]Col. George Everest (1790–1866) served as Surveyor General in India from 1830 to 1843. *Mount Everest*, the highest on Earth, and the highest peak in the Great Trigonometric Survey of India was renamed in his honour. I say renamed, as the mountain was (and still is) *Sagarmatha* in Nepali, and *Chomolungma* ('Mother Goddess of the Universe') in Tibetan.

location of Britain's highest mountain Ben Nevis (1,345 m). Let a baseline be chosen such that $\alpha = 30°$ and $\beta = 45°$. The baseline $a = 984.6$ m.

$$h = \frac{a \tan \alpha \tan \beta}{\tan \beta - \tan \alpha} \qquad (10.9)$$

$$\therefore h = 984.6\text{m} \times \frac{\tan 30° \tan 45°}{\tan 45° - \tan 30°} = 1{,}345\,\text{m} \qquad (10.10)$$

From the top of Ben Nevis the horizon (which can be seen on a clear day as one gazes East towards the Small Isles and Outer Hebrides beyond, or perhaps more easily as one looks south-west down Loch Linnhe towards Oban, as illustrated in Fig. 10.1) should have a dip angle of about $\gamma \approx 1.2°$. Al-Biruni's equation therefore predicts the radius of the Earth to be:

$$R = \frac{h \cos \gamma}{1 - \cos \gamma} \qquad (10.11)$$

$$R = \frac{1{,}345\text{m} \times \cos 1.2°}{1 - \cos 1.2°} \approx 6.13 \times 10^6 \text{ m} \qquad (10.12)$$

This is an error of about:

$$\frac{R_\oplus - R}{R_\oplus} = \frac{6.371 - 6.13}{6.371} \approx 4\% \qquad (10.13)$$

To achieve the exact radius of $R_\oplus = 6.371 \times 10^6$ m from an altitude of $h = 1{,}345$ m, the dip angle is:

$$\gamma = \cos^{-1}\left(\frac{1}{1 + \frac{h}{R_\oplus}}\right) = 1.177° \qquad (10.14)$$

10.3 Solving the Horizon Problem

The geometry used by Al-Biruni to measure the radius of the Earth R can also be applied to determine the distance d to the horizon, for an observer at altitude h above the Earth's surface. Instead of using trigonometry, we can use Pythagoras' theorem applied to the sides of Al-Biruni's horizon triangle in Fig. 10.1(c):

$$(R + h)^2 = R^2 + d^2 \qquad (10.15)$$

$$R^2 + 2Rh + h^2 = R^2 + d^2$$

$$2Rh\left(1 + \frac{h^2}{2Rh}\right) = d^2$$

$$\sqrt{2Rh}\left(1 + \frac{h}{2R}\right)^{\frac{1}{2}} = d \qquad (10.16)$$

Now the altitude h is clearly going to be much less than the radius of the Earth R, so we can use a Binomial Expansion since $\frac{h}{2R} \ll 1$:

$$d = \sqrt{2Rh} \left(1 + \frac{h}{4R} + \tfrac{1}{2}\left(-\tfrac{1}{2}\right)\left(\frac{h}{2R}\right)^2 + \cdots \right) \tag{10.17}$$

If we take the crudest approximation:

$$d \approx \sqrt{2Rh} \tag{10.18}$$

Using $R_\oplus = 6.371 \times 10^6$ m:

$$d \approx 3.57\,\text{km} \times \sqrt{h} \tag{10.19}$$

where h is in metres. This means at an altitude of $h = 100$ m, the horizon is about 35.7 km away. Note the refraction of light by the atmosphere will modify the distance one could potentially see, since straight-line rays will be curved by a varying refractive index with air density and temperature. This a particularly noticeable effect in microwave[5] propagation, which is the type of electromagnetic waves used by radar systems. It is typical in radar engineering to use the 'four-thirds Earth' model [16] to account for refraction in distance-to-the-horizon calculations. In other words, $R \longrightarrow \frac{4}{3}R$ in the equation $d \approx \sqrt{2Rh}$. Hence the microwave horizon is about $d \approx \frac{2}{\sqrt{3}} \times 3.57\,\text{km} \times \sqrt{h} \approx 4.12\,\text{km} \times \sqrt{h}$. A radar on the top of a 20-m mast of a ship should therefore be able to see surface targets up to a range of about 18.4 km.

10.4 A Choice of Coordinates to Navigate the Sphere

10.4.1 *Earth centred GEN and spherical polars (range, latitude and longitude)*

In order to plot a trajectory on, or beyond the surface of a spherical Earth using a computer graphics package, you will typically require a set of *Cartesian* x, y, z style *orthonormal*[6] coordinates for points on the trajectory. These coordinates might sensibly originate at the centre of the Earth. In this chapter, we shall refer to this most fundamental *geocentric* coordinate set as $\mathbf{G}, \mathbf{E}, \mathbf{N}$, i.e. a set of basis vectors that point towards the **G**reenwich meridian, **E**ast and **N**orth from the Earth's geometric centre. $\mathbf{G}, \mathbf{E}, \mathbf{N}$ coordinates have the properties:

$$|\mathbf{G}| = |\mathbf{E}| = |\mathbf{N}| = 1 \tag{10.20}$$

$$\mathbf{G} \cdot \mathbf{E} = \mathbf{G} \cdot \mathbf{N} = \mathbf{E} \cdot \mathbf{N} = 0 \tag{10.21}$$

---

[5]That is, frequencies between about 30 MHz to about 10 GHz.

[6]*Orthonormal* means the set of three vectors which define the coordinate system are mutually perpendicular, and each have a magnitude of unity.

$$\mathbf{G} \times \mathbf{E} = \mathbf{N} \tag{10.22}$$

$$\mathbf{E} \times \mathbf{N} = \mathbf{G} \tag{10.23}$$

$$\mathbf{N} \times \mathbf{G} = \mathbf{E} \tag{10.24}$$

The latter set of cross-products also define $\mathbf{G}, \mathbf{E}, \mathbf{N}$ as a 'right-handed-set'. The problem with $\mathbf{G}, \mathbf{E}, \mathbf{N}$ is that most trajectories that have significance to the Earth are those which incorporate its spherical symmetry, for example, movement on the surface. A *spherical polar* coordinate set is far more suitable, and we shall define it in terms of range R, latitude λ and longitude ϕ as illustrated in Fig. 10.2. If a position vector from the Earth centre is $\mathbf{r} = X\mathbf{G} + Y\mathbf{E} + Z\mathbf{N}$ (i.e. (X, Y, Z) in $\mathbf{G}, \mathbf{E}, \mathbf{N}$ coordinates):

$$X = R \cos \lambda \cos \phi \tag{10.25}$$

$$Y = R \cos \lambda \sin \phi \tag{10.26}$$

$$Z = R \sin \lambda \tag{10.27}$$

10.4.2 *Drawing circles on a sphere*

Pretty much all the classic problems of navigation using spherical geometry reduce to the ability to draw a circle on the surface of a sphere, as illustrated in Fig. 10.3. The most obvious in digital map-making might be how to draw lines of latitude and longitude at say $10°$ intervals. What are the X, Y, Z values in $\mathbf{G}, \mathbf{E}, \mathbf{N}$ coordinates? In logistics, the first approximation to long-distance transport route planning via plane or ship is to consider the minor arc of a *great circle* between two points on the Earth's surface defined by their latitude and longitude. This the shortest distance, on the spherical surface, between these points.[7] In Fig. 10.4, the shortest surface journey (about 9,548 km[8]) between London and Hong Kong is calculated.

A generic sphere-surface circle shall be defined by firstly defining a local Cartesian x, y, z coordinate system based upon orthonormal unit vectors $\{\widehat{\mathbf{x}}, \widehat{\mathbf{y}}, \widehat{\mathbf{z}}\}$, defined in a plane tangent to the surface of a sphere at range R and latitude λ and longitude ϕ. We shall refer to this orthonormal, right-handed coordinate system as 'East, North, Up', as the y axis shall point towards the geometric North pole and the z axis shall

---

[7]Obviously in reality the impact of surface currents, weather and the presence of land will result in a significant departure from the great circle ideal in a maritime scenario. Great circle trajectories for commercial aircraft journeys are perhaps more accurate representations, although obviously the effect of the Earth's rotation, high altitude airflows such as the jet-stream, plus human obstacles such as no-fly-zones, will result in significant modification.

[8]The London to Hong-Kong calculation of 9,548 km, based upon a spherical Earth and between city centres, is very close to the Google Maps estimate of 9,617 km. One assumes the latter should be more accurate, perhaps to the nearest aircraft runway!

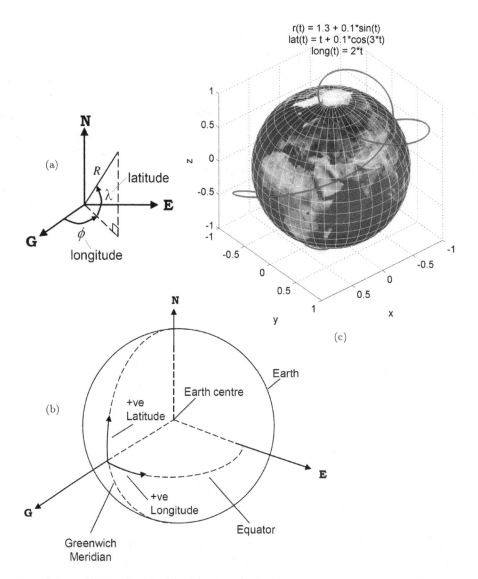

Fig. 10.2   Geometric relationship (a), (b) between Earth-centered Cartesian coordinates (**North**, **E**ast, **G**reenwich) and spherical polar coordinates range R, longitude ϕ and latitude λ. Interesting 3D trajectories (e.g. satellite orbits, etc.) can be plotted (c) by defining R, ϕ, λ trajectories *parametrically* in terms of periodic functions. Earth texture from Wikipedia: https://commons.wikimedia. org/wiki/File:Earthmap1000x500.jpg.

be on a radial from the Earth's centre, i.e. 'up'. In terms of $\mathbf{G}, \mathbf{E}, \mathbf{N}$ basis vectors:

$$\widehat{\mathbf{z}} = (\mathbf{G}\cos\phi + \mathbf{E}\sin\phi)\cos\lambda + \mathbf{N}\sin\lambda \qquad (10.28)$$

$$\widehat{\mathbf{y}} = -(\mathbf{G}\cos\phi + \mathbf{E}\sin\phi)\sin\lambda + \mathbf{N}\cos\lambda \qquad (10.29)$$

$$\widehat{\mathbf{x}} = \widehat{\mathbf{y}} \times \widehat{\mathbf{z}} = \mathbf{E}\cos\phi - \mathbf{G}\sin\phi \qquad (10.30)$$

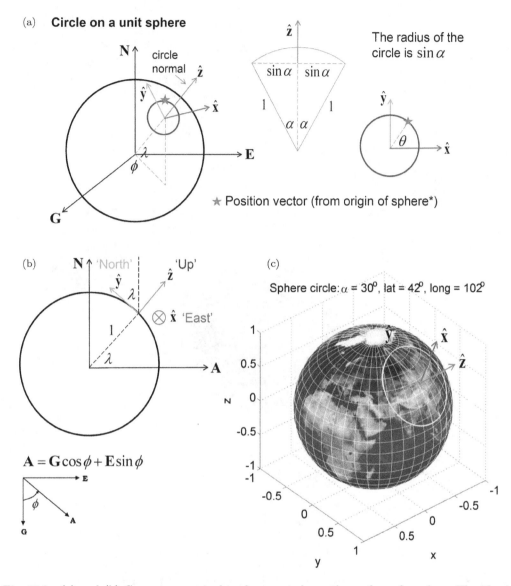

Fig. 10.3 (a) and (b) Geometry required to draw a circle on the surface of a sphere. The idea is to define a circle in terms of a Cartesian $\hat{\mathbf{x}}, \hat{\mathbf{y}}, \hat{\mathbf{z}}$ basis vector set with the $\hat{\mathbf{z}}$ axis pointing radially out from a spherical surface, through the (latitude, longitude) of a point of interest, and extending from the centre of the circle. If the $\hat{\mathbf{x}}, \hat{\mathbf{y}}, \hat{\mathbf{z}}$ coordinates can be defined in terms of Earth-centered North, East, Greenwich Cartesians, the circle can be easily plotted using standard computer-based 3D plotting tools. As indicated in (c), as well as drawing the yellow circle, the technique can be used to draw lines of latitude and longitude too. Earth texture from Wikipedia: https://commons. wikimedia.org/wiki/File:Earthmap1000x500.jpg.

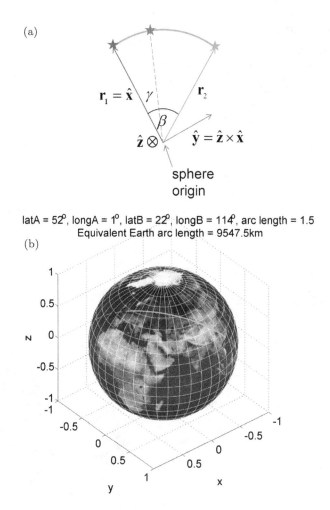

(a)

$\mathbf{r}_1 = \hat{\mathbf{x}}$ $\quad \gamma$ $\quad \mathbf{r}_2$

β

$\hat{\mathbf{z}} \otimes$ $\quad \hat{\mathbf{y}} = \hat{\mathbf{z}} \times \hat{\mathbf{x}}$

sphere
origin

latA = 52°, longA = 1°, latB = 22°, longB = 114°, arc length = 1.5
Equivalent Earth arc length = 9547.5km

(b)

Fig. 10.4 The circle-on-a-sphere idea can be used to plot *great circles* (i.e. Earth radius sized circles) that have a minor arc which passes through two points on the surface of the Earth. These are represented by the red and green stars in (a). The minor arc is plotted between them in (a), and overlaid upon an Earth texture in (b). This great circle computation is very useful for calculating the ground-distance between cities, and a first approximation to aircraft time-of-flight calculations. (It might not actually be the optimal trajectory, given the effect of wind and also the 'Coriolis force', an effect which manifests because the Earth itself is rotating and therefore every point on the surface that is not on the rotation axis is accelerating). Earth texture from Wikipedia: https://commons.wikimedia.org/wiki/File:Earthmap1000x500.jpg.

As illustrated in Fig. 10.3, these can be readily inferred from a pair of 2D diagrams, using unit vector $\mathbf{A}$ drawn in the equatorial plane (i.e. at zero latitude) at longitude ϕ between $\mathbf{G}$ and $\mathbf{E}$ vectors.

$$\mathbf{A} = \mathbf{G}\cos\phi + \mathbf{E}\sin\phi \tag{10.31}$$

$$\hat{\mathbf{z}} = \mathbf{A}\cos\lambda + \mathbf{N}\sin\lambda \tag{10.32}$$

$$\hat{\mathbf{y}} = -\mathbf{A}\sin\lambda + \mathbf{N}\cos\lambda \tag{10.33}$$

To define and plot a trajectory based upon (x, y, z) coordinates, we can use the position vector defined via $\{\mathbf{G}, \mathbf{E}, \mathbf{N}\}$ and (R, λ, ϕ) as:

$$\mathbf{r} = R \cos \lambda \cos \phi \mathbf{G} + R \cos \lambda \sin \phi \mathbf{E} + R \sin \lambda \mathbf{N} + x\widehat{\mathbf{x}} + y\widehat{\mathbf{y}} + z\widehat{\mathbf{z}} \qquad (10.34)$$

We now have the tools to define a sphere-surface circle drawn on a suitable plane characterised by (R, λ, ϕ) and $\{\widehat{\mathbf{x}}, \widehat{\mathbf{y}}, \widehat{\mathbf{z}}\}$, and hence in terms of geocentric Cartesians $\{\mathbf{G}, \mathbf{E}, \mathbf{N}\}$. The trick is to use a tangent plane that is moved *below* the Earth's surface by $R(1 - \cos \alpha)$ where $r = R \sin \alpha$ is the radius of the circle. Perhaps an easier way of thinking about this is to translate the $\{\widehat{\mathbf{x}}, \widehat{\mathbf{y}}, \widehat{\mathbf{z}}\}$ tangent plane to the centre of the Earth, and define the position vector $\mathbf{r}$ of the circle from this point.

$$\mathbf{r} = R \sin \alpha \left(\widehat{\mathbf{x}} \cos \theta + \widehat{\mathbf{y}} \sin \theta \right) + \widehat{\mathbf{z}} R \cos \alpha \qquad (10.35)$$

The angle θ is the polar angle of the circle, defined anticlockwise from the $\widehat{\mathbf{x}}$ axis in the $\widehat{\mathbf{x}}, \widehat{\mathbf{y}}$ plane. To plot a circle using MATLAB, we would define a linearly spaced vector θ angles between 0 and 2π radians in say 300 steps, and then use the equations above to determine $\mathbf{r}$ in X, Y, Z coordinates, that a 3D plotting function `plot3()` will accept.

Lines of longitude, latitude and (generic) great circles are special cases:

$$\begin{aligned} \text{Lines of longitude} \quad & \alpha = \frac{\pi}{2}, \lambda = 0 \\[2mm] \text{Lines of latitude} \quad & 0 < \alpha \le \frac{\pi}{2}, \lambda = \pm \frac{\pi}{2} \\[2mm] \text{Any great circle} \quad & \alpha = \frac{\pi}{2} \end{aligned} \qquad (10.36)$$

Note if at first glance the values quoted for λ seem strange for lines of latitude and longitude, remember the values here are for the $\widehat{\mathbf{x}}$, $\widehat{\mathbf{y}}$, $\widehat{\mathbf{z}}$ unit vectors of the tangent plane, in which we define in circle. In other words, λ, α define the orientation of the circle tangent plane, not the points on the circle.

10.4.3 *Great circle minor arc between two points on a sphere*

The arc (see Fig. 10.4) between two locations on the surface of a sphere of radius R, defined by positions vectors $\mathbf{r}_1$, $\mathbf{r}_2$ from the centre of the sphere, has angle β radians, which can be found from the dot product of the position vectors:

$$\mathbf{r}_1 \cdot \mathbf{r}_2 = |\mathbf{r}_1| |\mathbf{r}_2| \cos \beta \qquad (10.37)$$

$$\therefore \beta = \cos^{-1} \left(\frac{\mathbf{r}_1 \cdot \mathbf{r}_2}{|\mathbf{r}_1| |\mathbf{r}_2|} \right) \qquad (10.38)$$

The minor arc is the minimum of β and $2\pi - \beta$. Unless the two points are diametrically opposite, we can form a unique normal vector $\widehat{\mathbf{z}}$ to the great circle which passes through the two locations.[9]

$$\widehat{\mathbf{z}} = \frac{\mathbf{r}_1 \times \mathbf{r}_2}{|\mathbf{r}_1 \times \mathbf{r}_2|} \tag{10.39}$$

If we define $\widehat{\mathbf{x}} = \frac{\mathbf{r}_1}{|\mathbf{r}_1|}$, and therefore to form a right handed set:

$$\widehat{\mathbf{y}} = \widehat{\mathbf{z}} \times \widehat{\mathbf{x}} \tag{10.40}$$

The position vector of the arc (from the centre of the Earth) is therefore parameterised by angle γ (which as in the previous section could be defined as a linearly spaced vector in say 300 steps for a smooth looking curve):

$$\mathbf{r} = R(\widehat{\mathbf{x}} \cos \gamma + \widehat{\mathbf{y}} \sin \gamma) \tag{10.41}$$

$$0 \leq \gamma \leq \beta \tag{10.42}$$

10.4.4 The 'Death Star' problem – defining a parabolic indent or cap on a sphere

The moon of Saturn, *Mimas* (diameter 396 km) has a huge eyeball shaped impact crater, which makes it look somewhat like the *Death Star* in the *Star Wars* films. A parabolic dip (or indeed Pinnochio-like 'nose', see Fig. 10.5) can be drawn on a sphere of radius R defined from a grid of X, Y, Z coordinates, which are themselves defined from a mesh of latitude and longitude values. In MATLAB, the process of plotting a spherical surface of radius R would be via the following syntax[10]:

```
R = 1.23; lat = linspace( -pi/2, pi/2, 500 );
long = linspace( 0, 2*pi, 1000 );
[lat, long] = meshgrid( lat, long );
X = R*cos(lat).*cos(long);
Y = R*cos(lat).*sin(long);
Z = R*sin(lat);
surf( X,Y,Z ); axis vis3d; shading interp;
```

Let:

$$\widehat{\mathbf{z}} = \cos \lambda \cos \phi \mathbf{G} + \cos \lambda \sin \phi \mathbf{E} + \sin \lambda \mathbf{N} \tag{10.43}$$

be a unit radial vector defining the centre line of the parabolic indent. Define an indent ΔR which occurs over angular range $-\alpha \leq \eta \leq \alpha$:

$$\Delta R = R \times k \left(\frac{\sin^2 \eta}{\sin^2 \alpha} - 1 \right) \tag{10.44}$$

---

[9]Note if the locations are on a diameter (i.e. $\beta = \pi$) there is an infinite variety of possible great circles that can pass through the points.

[10]The .* notation is used to ensure matrix multiplication is element-wise. Note after the meshgrid operation, lat, long are now matrices of dimensions 500×1000.

(a) Death Star: $\alpha = 15^\circ$, lat = 21.2°, long = -63°, k = 0.2

(b) normal line

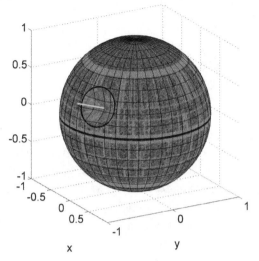

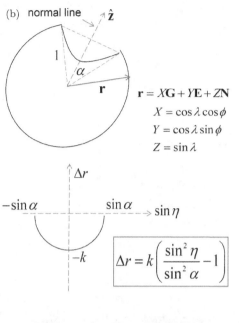

$\mathbf{r} = X\mathbf{G} + Y\mathbf{E} + Z\mathbf{N}$

$X = \cos\lambda\cos\phi$

$Y = \cos\lambda\sin\phi$

$Z = \sin\lambda$

$$\Delta r = k\left(\frac{\sin^2\eta}{\sin^2\alpha} - 1\right)$$

(c) Death Star: $\alpha = 15^\circ$, lat = -219°, long = 195°, k = -0.75

(d)

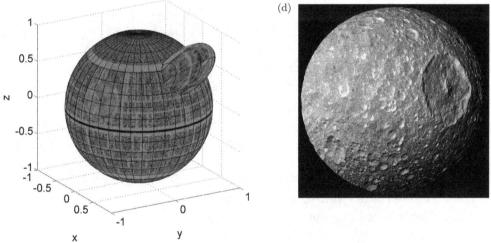

Fig. 10.5 A playful implementation of circle-on-a-sphere drawing, and also the idea of a parabolic indent. The surface deviation $\Delta R = R \times k\left(\frac{\sin^2\eta}{\sin^2\alpha} - 1\right)$, defined in (b) for a sphere of radius R, can be used to construct the antenna on a 3D model of the Star-Wars *Death Star* in (a). The antenna is the one which focuses the planet-exploding green beam of doom. Note the moon of Saturn, *Mimas* in (d) has as similarly iconic 'eye'. If the parabola k parameter is inverted (c), the indent becomes a giant nose! Mimas image source: NASA/JPL-Caltech/Space Science Institute. Wikipedia: Mimas (moon).

A positive k defines an indent, whereas a negative k defines a nose-like protrusion. The idea is to find the subset of position vectors $\mathbf{r}$ which satisfy $|\eta| \leq \alpha$ where:

$$\eta = \cos^{-1}\left(\frac{\mathbf{r} \cdot \widehat{\mathbf{z}}}{R}\right) \tag{10.45}$$

This subset of position vectors are then scaled by factor $1 + k\left(\frac{\sin^2 \eta}{\sin^2 \alpha} - 1\right)$.

10.5 Why Navigate the Sphere?

The genesis of this chapter was a delightful encounter with Dr Chas McCaw, Head of Science at Winchester College. He told me he had been researching the mathematics of navigation, and associated spherical geometry, that would have been essential knowledge to seafaring nations hundreds of years ago. The nature and cultural aspects of this ancient topic often resulted in deeply complicated analysis, all couched in esoteric language. These days, satellite positioning systems, jet engines and diesel-powered shipping make such knowledge largely redundant, and certainly all but the simplest geometric concepts have been weeded out of most mathematics and physics courses in the UK. However, I think the historical context of navigation, starting with Eratosthenes and Al-Biruni's calculation of the radius of the Earth, is a fascinating gateway to learning even the most basic of trigonometrical ideas. The distance to the horizon remains an important range limit for remote sensing with line-of-sight microwave systems such as radar,[11] and is pretty interesting anyway, particularly if you spent your childhood staring at the horizon from a beach or clifftop. Plotting circles on spheres is a modern way of solving nearly all the historical navigation problems, and is an incentive to learn about different three-dimensional coordinate systems and how to convert between them. If terrestrial navigation problems are too tame, the same mathematics is needed if you wish to track orbiting spacecraft. The spectacular success of recent space missions, from the Mars Rovers, to Juno & Cassini, to New Horizons, to the SpaceX Dragon capsule, are no stroke of luck. Each mission is entirely dependent on a precise knowledge of both the position and orientation of these extraterrestrial sensors, minuscule objects in the vastness of space.

---

[11]'High-frequency' radar (2–30 MHz typically) which perversely is at the lower end of radar frequencies, can actually make use of propagation *over the horizon*. There are two main mechanisms: 'skywaves' at these frequencies bounce off the ionosphere, and can therefore travel huge distances before attenuating beyond the noise threshold of a radar receiver. 'Ground waves' (which should really be called *ocean waves*) propagate around the curvature of the Earth, making use of the conductivity of sea-water.

Chapter 11

Modelling Money and Mortgages

Capital, cash, bread, bucks, funds, finances, wads, wonga. This chapter is about aspects of *money*, with particular focus on simple models of a few important economic concepts that most people will have to actively, or indirectly, utilise during their lives:

- The concept of a **mortgage**, i.e. a loan to provide sufficient capital to enable the purchase of a domestic property, payable over a long-time period (e.g. twenty to forty years). We will develop a mathematical model of the most common variant, the 'fixed rate repayment mortgage'. This is where a borrower agrees to pay back a fixed monthly payment p over N years,[1] and after this time, owns the house and has nothing more to pay to the lender. The mortgage is characterised by an annual % rate of interest R, in addition to p and N. The generational divide between 'early retired and mortgage free' and 'work and rent until you die' is currently one of the major social problems in the UK, largely driven by a huge rise in house prices since the late 1980s.[2]

- A model of a long-term **savings scheme**, based upon a fixed amount p saved monthly, and an (annual) rate of interest R. At the time of writing, R in the UK is not much more than 0%. So, this chapter can be regarded as rose-tinted fiction if you are a saver, or a foolish harbinger of fiscal doom if you are currently living with the burden of significant debt.

- A model of a **share trading scheme**, an enormously simplified version of what fund-managers of investment portfolios and pensions might do on a given day

---

[1]In most cases fixed rate mortgages are only really fixed for a few years, before having to be renegotiated. As of Jan 2019, a quick scan of a website like www.money.co.uk shows all the fixed rate mortgages offer an interest rate of about 2% or lower for 2 years, before reverting to an annual interest rate of about 4.2%.

[2]https://www.economicshelp.org/blog/8733/housing/uk-house-prices-high/ (accessed 14/4/2020).

of trading. But I'm guessing here. I'm sure it involves a combination of voodoo, coin flipping, divination and magic.

11.1 Context: A British Housing Crisis

There is currently a housing crisis in Britain, and it appears to be getting worse. Liam Geraghty of the homeless charity *The Big Issue* spelled out the problem in 2018[3]:

- The National Housing Federation calls for 340,000 new homes to be built each year. In 2017 only 217,350 were constructed.[4] There remains a significant shortfall, and particularly for homes deemed 'affordable' and available for social rent and shared ownership. The discrepancy between supply and demand will therefore continue to drive up prices in an economic model without imposed price increase limits. According to the UK Housing Index, the annual price growth (2018) is 4.2%, with an average house price of £224,144. According to the Office of National Statistics (ONS), this is 7.8 times the average salary, which makes the cost of mortgages prohibitive too all but the most wealthy. The variation in UK house prices from Jan 2005 till Nov 2018 is plotted in Fig. 11.1.
- Over 200,000 empty or second homes are left dormant for over six months, further compounding the housing supply problem.[5]
- The lack of affordable housing stock, and high prices, are forcing many to rent. The private rental market has 'more than doubled' in the past 20 years, with about 20% of all residential properties (about 4.9 million homes) being rented. As per the house purchase market, limited supply and high demand have driven up prices, with average rents in London costing about 49% of average salaries.[6] One assumes the additional costs of a rental property incurred by a landlord

---

[3]https://www.bigissue.com/latest/why-does-the-uk-have-a-housing-crisis/ (accessed 31/12/2018).
[4]https://www.telegraph.co.uk/business/2017/11/16/government-celebrates-building-217000-homes-year-hitting-target/.
[5]There is a broader social impact too. Empty properties in idyllic rural locations, that are unaffordable to locals desperately trying to get on the 'housing ladder', is a somewhat galling state of affairs, and can lead to a sad 'ghost town' phenomenon. Many coastal regions of the UK such as Cornwall are stunningly beautiful, and attract swarms of metropolitan second home owners. Sadly the local population are subject to some of the highest levels of poverty in Britain.
[6]The 49% figure was quoted by Geraghty in December 2018. Clearly there will be huge local (and personal) variations, but given the rise in house prices relative to wage stagnation (or significant reduction as a result of the 2020–2021 pandemic), one could easily imagine a much higher figure, particularly if one factors other basic living costs such as council tax, utilities, insurance etc. While it is clearly unreasonable to describe comfortably accommodated First-World professional adults as experiencing poverty in any real way, it is nonetheless a growing likelihood that income only just about exceeding basic costs, is becoming a harsh reality for a growing fraction of the UK population.

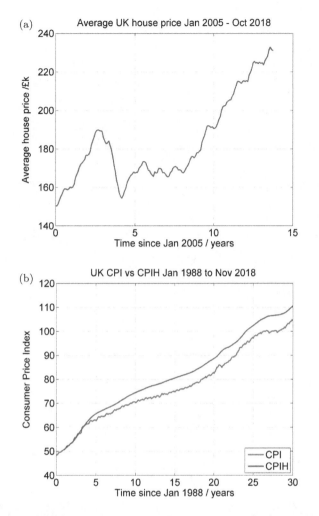

Fig. 11.1 (a) Average UK house price since January 2005. A dramatic increase, and the cause of much fortune and misfortune, depending on when you wished to purchase, or sell, a residential property. *Source*: www.ons.gov.uk/economy/inflationandpriceindices/bulletins/ housepriceindex/october2018. (b) Graph of UK Consumer Price Index (CPI) vs time. This is, in essence, the cumulative effect of inflation on the value of money. About £48 in January 1988 is worth about £105 in January 2018. For any investment to be profitable, be it property, a savings scheme or stock market speculation, the gain in value must exceed the CPI curve. CPI is calculated by the Bank of England and takes into account the prices of a variety of commodities. The CPIH curve takes into account the value of housing, in addition to items associated with the CPI calculation. *Source*: https://www.ons.gov.uk/economy/inflationandpriceindices/timeseries/l55o/mm23.

(e.g. increased council taxes, higher buy-to-let mortgages) are passed onto those renting, further pushing up prices.

Overall the picture is pretty bleak for those starting out on a career. Low interest rates in the UK (0.1% as of February 2021) and the concerns about the impact of

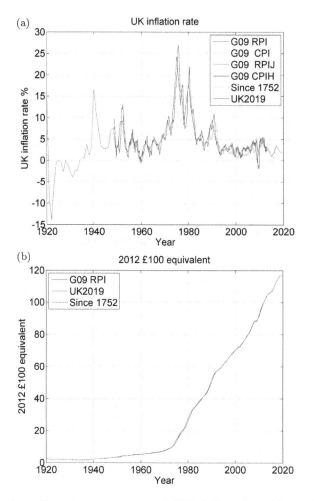

Fig. 11.2 (a) Overlay of various measures of UK inflation such as Retail Price Index (RPI), Consumer Price Index (CPI) and CPI + Housing (CPIH). *Source*: *Guardian Datablog* 2009. The UK Composite Index since 1752 is from: https://www.officialdata.org/articles/uk-composite-price-index-values-since-1750/. Graph (b) is the equivalent value of £100 in 2012, perhaps a more useful measure of the cumulative effect of inflation. £10 in the late seventies is worth about £120 in 2020. This is a useful calculation when calming down elderly relatives when they baulk at modern restaurant prices, but a depressing calculation if your salary and savings have been frozen for years.

'Brexit' compound the problem of saving for a house. Given inflation rates (see Fig. 11.2) are about 0.9% (as of February 2021[7]), this means savers are effectively losing money, while house prices continue to grow.

The author has a rather pressing vested interest in this particular problem being solved (!), but I shall leave the pontifications and interventions to the politicians

---

[7]https://www.ons.gov.uk/economy/inflationandpriceindices (accessed 19/2/2021).

and economists. The purpose of this chapter is to develop a basic set of tools to enable mathematical models to be constructed of (i) savings schemes, (ii) fixed-rate mortgages and (iii) share trading. Although the present economic climate in the UK may not be conducive to enable a house purchase for many, at least a rational understanding of the basic associated finances may allow for sensible planning and avoidance of crippling debt. Ignorance is never bliss, particularly when it comes to personal financial matters. Alternatively, as the rather pessimistic Private Frazer, played by Scottish actor John Laurie, often said in *Dad's Army*; a clear-eyed fiscal analysis might simply reveal that indeed "we're doomed...!"

11.2 Fixed Rate Mortgages

Everybody needs somewhere to live. This is perhaps one of the fundamental problems associated with a relatively 'free' housing market (i.e. one with limited governmental intervention), since unlike other similarly expensive commodities like fine art, yachts and sports cars, there is very significant pressure to buy into the housing market. At present, one might conclude that this gives landlords and house builders rather too much power over buyers, especially first-time buyers. The vast majority of people have an immediate need to buy or rent a property before they have a chance to save up enough to purchase one outright. If we stick to complete house purchases for simplicity, this means borrowing money from a financial institution in the form of a long-term loan, i.e. a *mortgage*. There are a huge variety of mortgages, for example *interest-only*,[8] *trackers*,[9] *Australian*,[10] to name but a few. We shall construct a model of the simplest 'vanilla' mortgage, which is where a borrower pays a fixed monthly payment p over N years until the mortgage is paid off. The money paid during the N years covers both the initial loan M_0 and a fixed annual % rate of interest R.

For brevity, let us define a *monthly interest fraction* k in terms of the annual percentage rate R:

$$k = \tfrac{1}{12} \times \tfrac{1}{100} R \tag{11.1}$$

---

[8]The mortgage payments only cover the interest on the loan, i.e. at the end of the mortgage term, the bank still owns 100% of the house.

[9]Mortgage payments vary according to the current interest rate. This can be a problem for long-term financial planning if there is a significant interest rate rise. A 'just manageable' tracker mortgage at a time of historically low interest rates may be a rather risky prospect.

[10]This is a mortgage which allows borrowers to overpay to reduce interest changes. In some schemes the mortgage is, in essence, a huge overdraft (i.e. a negative bank balance), which can be increased or reduced flexibly over time.

e.g. if $R = 5\%$, $k = \frac{1}{240}$. The amount still owed M_1 after one month of the mortgage is kM_0 of interest, minus p paid off. Hence:

$$M_1 = M_0 + kM_0 - p$$
$$M_1 = M_0 (1 + k) - p \tag{11.2}$$

A similar result can therefore be stated for the total amount still owed after 2 months:

$$M_2 = M_1 (1 + k) - p$$
$$M_2 = M_0 (1 + k)^2 - p (1 + k) - p \tag{11.3}$$

And 3 months...

$$M_3 = M_2 (1 + k) - p$$
$$M_3 = M_0 (1 + k)^3 - p (1 + k)^2 - p (1 + k) - p \tag{11.4}$$

i.e. an iterative relation:

$$M_{n+1} = M_n (1 + k) - p \tag{11.5}$$

This can be written out in full as:

$$M_n = M_0 (1 + k)^n - p \sum_{j=1}^{n} (1 + k)^{j-1} \tag{11.6}$$

The summation term is a *Geometric Progression* which can be evaluated as:

$$\sum_{j=1}^{n} x^{j-1} = \frac{1 - x^n}{1 - x} \tag{11.7}$$

Hence:

$$M_n = M_0 (1 + k)^n - p \frac{1 - (1 + k)^n}{1 - (1 + k)}$$
$$M_n = M_0 (1 + k)^n + \frac{p}{k} (1 - (1 + k)^n)$$
$$M_n = M_0 (1 + k)^n - \frac{p}{k} ((1 + k)^n - 1) \tag{11.8}$$

The mortgage is paid off after $n = 12N$ months, i.e.

$$M_{12N} = 0 \tag{11.9}$$

Hence:

$$0 = M_0 \left(1 + k\right)^{12N} - \frac{p}{k} \left(\left(1 + k\right)^{12N} - 1 \right)$$

$$\therefore p = k \frac{M_0 \left(1 + k\right)^{12N}}{\left(1 + k\right)^{12N} - 1}$$

$$\therefore p = \frac{kM_0}{1 - \left(1 + k\right)^{-12N}} \tag{11.10}$$

The mortgage amount still owed after n months is therefore:

$$M_n = \left(1 + k\right)^n \left(M_0 - \frac{p}{k} \right) + \frac{p}{k} \tag{11.11}$$

$$p = \frac{kM_0}{1 - \left(1 + k\right)^{-12N}} \tag{11.12}$$

$$k = \tfrac{1}{1200} R \tag{11.13}$$

which, substituting for p, can be written as:

$$\frac{M_n}{M_0} = \left(1 + k\right)^n \left(1 - \frac{1}{1 - \left(1 + k\right)^{-12N}} \right) + \frac{1}{1 - \left(1 + k\right)^{-12N}}$$

$$\frac{M_n}{M_0} = \frac{1 - \left(1 + k\right)^n \left(1 + k\right)^{-12N}}{1 - \left(1 + k\right)^{-12N}} \tag{11.14}$$

$$\therefore \frac{M_n}{M_0} = \frac{\left(1 + k\right)^{12N} - \left(1 + k\right)^n}{\left(1 + k\right)^{12N} - 1} \tag{11.15}$$

The total amount paid is $12Np$, which means the total amount of interest paid on the original loan of M_0 is:

$$I = \frac{12NkM_0}{1 - \left(1 + k\right)^{-12N}} - M_0 \tag{11.16}$$

A useful indicative measure is therefore the ratio of interest I to initial mortgage value M_0:

$$\Gamma = \frac{I}{M_0} = \frac{12Nk}{1 - \left(1 + k\right)^{-12N}} - 1 \tag{11.17}$$

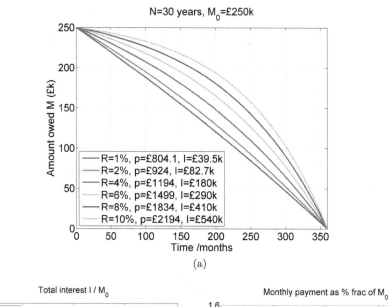

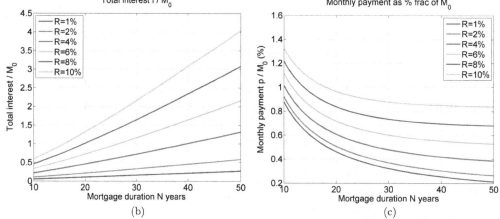

Fig. 11.3   Model of a fixed rate 'vanilla' repayment mortgage. This is a very common type of loan that enables money to be borrowed to finance the purchase of a house. Repayment is typically over many decades. Graph (a) tracks the value owed £M over the mortgage term. Depending on the fixed annual percentage interest rate R, the fixed monthly payment p varies, as does the curvature of the repayment curve. (b) A repayment mortgage can result in a significant multiple of the amount borrowed £M_0, being paid over the mortgage term. For a 5%, 30-year mortgage, your interest will be about the same as the money borrowed. (c) R can be a little misleading when it comes to calculating the fixed monthly payments, which are characteristic of this type of mortgage. The monthly payment p, calculated as a fraction of the amount borrowed M_0, decays with the length of mortgage term N (years). However, the steepness of this drop depends on R. If the latter is higher, the reduction with N is significantly less.

Let's run the numbers for $M_0 = £250,000$, $N = 30$ years, $R = 5\%$:

$$k = \frac{5}{1200} = \frac{1}{240} \tag{11.18}$$

$$p = \frac{\frac{1}{240} \times £250,000}{1 - \left(1 + \frac{1}{240}\right)^{-12 \times 30}} = £1,342.05 \tag{11.19}$$

$$\Gamma = \frac{12 \times 30 \times \frac{1}{240}}{1 - \left(1 + \frac{1}{240}\right)^{-12 \times 30}} - 1 = 0.933 \tag{11.20}$$

i.e. a monthly payment that is $\frac{1342.05}{250,000} = 0.54\%$ of the amount borrowed. The total amount of interest paid is $0.933 \times £250,000 = £233,250$. This means the borrower will have to pay almost double the house price over the duration of the mortgage. However, 'conventional wisdom' states that the expected growth in house prices should more than compensate for this cost. According to the ONS, average UK house prices have risen from £150,633 in January 2005 to £231,095 in October 2018, i.e. about 53.4% growth in 13.8 years. On the basis of uncompounded 'simple' interest, this means about 3.9% added per year. Note the relative linearity of house price growth in Fig. 11.1 during the last 5 years gives some basis to this assertion. Therefore, over 30 years we might expect a growth of about 117%. In crude terms, this means a borrower has potentially gained about $117\% - 93\% = 24\%$ of M_0.

A significant house price growth is perhaps the basis of the 'housing ladder' idea, whereby a family can acquire a progression of increasingly larger (and more expensive) houses without a huge cash injection such an inheritance, lottery win, unearthing of Anglo-Saxon treasure hoard, etc. If the *equity*, i.e. the difference between selling cost (the *asset*) and the remaining mortgage (the *liability*), is a significant enough sum, then this can help fund the deposit for the new house, thereby reducing an otherwise unaffordable mortgage.

Figure 11.3 illustrates our fixed-rate mortgage model. (a) is a graph of amount owed during the mortgage term, for a variety of R values for a $N = 30$ year mortgage. (b) and (c) show the variation of total interest and monthly payment with both R and N.

11.3 Regular Monthly Payment Savings and Investment Accounts

The perspective of this chapter is unashamedly one of the saver and potential house-buyer, and the overall goal is to create models of how sufficient finance could be raised to fund a house purchase (or at very least, the deposit on a house, to enable an affordable mortgage). The traditional model of savings accounts taught in schools is typically based on compound interest added annually to a fixed deposit. This is a nice geometric progression, but sadly bares little relevance to most financial realities! If you have a large enough lump sum to buy a house, and you wish to buy a house, then there is little point keeping it in a savings account. Alternatively, if you own a house, are retired and wish to draw from a lump sum, then this amount is also unlikely to be a static quantity, given regular money will be paid out.[11] A sensible savings scheme model is therefore one which *factors in regular payments*. We shall use the same notation as in the fixed rate mortgage model in the previous section, but in this case p shall be a monthly payment into a savings scheme, tracked over N years. Obviously rates of interest vary, and if you have investments in *equities* (i.e. shares) in particular, market *volatility* will mean somewhat noisy traces of your savings account balance vs time, with the occasional worrying drop during recessions, hopefully more than compensated for by upward trends in more financially productive years. Note the idea of 'pound cost averaging' can help smooth these deviations. When share prices are low (e.g. in a recession), the regular investment p will buy more shares than when stock prices are riding high. In other words, the regular investment during a recession will become 'supercharged' in a recovery if 'recession-bought shares' become more valuable. This is countered of course by a lessened impact of regular investment during a recovery period. Modelling stock market rises and falls is best left to clairvoyants and hedge-fund managers, whereas I think it is not unreasonable to model a savings or investment scheme as one with a fixed annual % rate of interest R. It is important to correctly model how interest is typically calculated and paid into a savings scheme. Most accounts will calculate interest daily, based upon $\frac{R}{365}$ (%) of the balance of the account. This is accumulated into interest I, which is then paid (typically) at the *end* of each year. I is then reset for the next year. In other words, the benefit of compound interest (i.e. interest on interest) only occurs on an annual cycle, not within a year.[12]

We shall construct a model of the variation of the balance S (in £) of a savings account vs time t in years, over duration N years, given monthly investment p and annual % interest R. MATLAB code is below, but the syntax is fairly generic to other high level computer languages. The general idea is we step through an array

---

[11]This concept of drawing regular payments from a large lump sum is a DIY version of an *annuity*. A purchased annuity is a scheme that will pay a fixed income until death, and the lump sum that pays for it will factor in the likelihood of the buyer of the annuity living for N years after the annuity is purchased, and sufficient profit for the provider of the annuity. Most workers will pay into a *pension* scheme during their career, which will typically provide an annuity following retirement. Unlike a DIY version, employers *as well as employees* contribute to pensions, and pension contributions are often very tax efficient.

[12]There are accounts which pay interest monthly, although one assumes the interest rate they offer will be lower to offset the potential loss to the banks. 'The house always wins' is a sensible mantra in dealings with financial institutions!

of numbers $1 + 365N$ elements long, which represents the daily value of savings S and investment V. The investment is simply what has been paid in (i.e. the initial deposit plus p multiplied by the months elapsed). For clarity, I use integer counters m and y which are incremented by one on a daily basis, until they are reset. Once m exceeds $365/12$ (i.e. an average month length), m is reset to zero and p is added to S, and V. Once $y > 365$, this is then reset to zero and the accrued interest I is added to S. Overall growth % g shall be calculated from the percentage increase of S compared to V.

```
% Initialize value of savings account. Assume no withdrawals!
S = [S0,zeros(1,365*N)];
% Initialize value of investment (no interest).
V = [S0,zeros(1,365*N)];
% Initialize month and year daily counters.
m = 0; y = 0;
% Initialize annual interest (£).
I = 0;
% Step through time and (i) add monthly payments, (ii) calculate interest
% (iii) add interest at the end of each year.
for n=1:N*365
    % Add daily interest
    I = I + (0.01*R/365)*S(n);
    % Increment month and year counters, and check if a month or year has passed.
    % If so add payment, or interest respectively.
    m = m + 1; y = y + 1;
    if m > 365/12
        %Month has elapsed. Reset month counter and add regular saving p.
        m = 0;
        S(n+1) = S(n) + p; V(n+1) = V(n) + p;
    else
        S(n+1) = S(n); V(n+1) = V(n);
    end
    if y > 365
        %Year has passed. Reset year counter and add interest accrued I over the
        %year. Then reset I to zero.
        y = 0; S(n+1) = S(n+1) + I; I = 0;
    end
end
%Compute overall growth percentage.
g = 100*( S(end)-V(end) )/V(end);
```

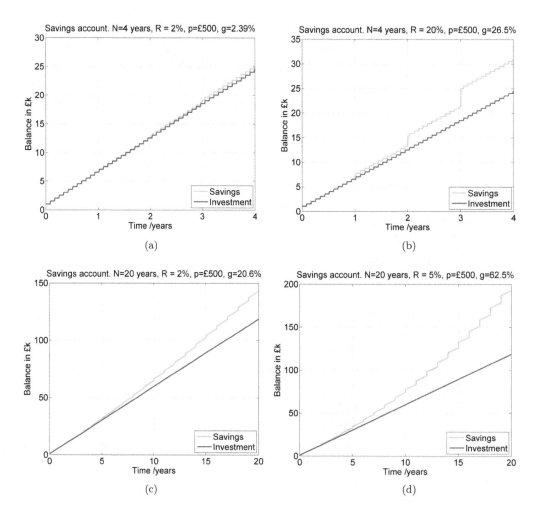

Fig. 11.4   A slightly more realistic model of a savings account. An initial deposit of £1,000 is made, and £500 of new money is saved monthly. The blue traces show the money invested, and the little vertical jumps correspond to the £500 additions. Assume annual compound interest (APR) at rate R % is calculated (and added) on an annual basis. The idea is that interest is calculated daily based upon $\frac{1}{365}R$ % of the total balance. This is tallied up but only actually added at the end of the year. You can see this represented by the larger vertical jumps in the green trace, which is the actual value of the savings account. Graphs (a) and (b) compare an APR of 2% to an APR of 20% for a savings account over a 4 year period. Graphs (c) and (d) compare a 2% and 5% APR over a 20-year period. However, at the time of writing (2021), UK interest rates are typically less than 1%, which would mean the green and blue curves would be almost indistinguishable.

The savings model is run in three illustrative scenarios, depicted in Fig. 11.4. Figure 11.4(b) represents a highly unrealistic $R = 20\%$ investment scheme, starting from an initial deposit of £1,000 and with £500 added each month. This is run for 4 years, with an overall growth of $g = 26.5\%$. The idea here is to clearly see the effect of annual interest being added. If such an investment scheme can be found, and is legal, do please email the author. Figures 11.4(a), 11.4(c) and 11.4(d) and perhaps present a more realistic scenario, over 4, and 20 years. The $R = 2\%$ scheme might be representative[13] of a cash-based savings account such as a Building Society *Individual Savings Account* (ISA), whereas the $R = 5\%$ might be the aspirational, and highly smoothed, average of a managed investment portfolio.

11.4 Save and Buy, or Buy, Rent, Sell, Buy?

As alluded to at the start of this chapter, the economic outlook for potential UK first-time house buyers is somewhat bleak, and appears to be getting worse. British society is very much historically geared to house ownership, and one can get a sense of personal failure if you live in a rented house. You will probably hear something along the lines of 'renting is dead money' from your house-owning parents and grand-parents, who tend to regard any other option as some form of heresy. Although I'm sure there are many fantastic socially-motivated landlords, the fact of the matter is that a renter will have significantly less power over their domestic arrangements, be it the rent, be it the ability to tailor a property to emerging needs, and even whether they can remain in the property in the long term. However, the currently very high ratio of house prices to wages means a mortgage is beyond an increasingly large sector of society in the UK. Demand for houses currently outstrips supply, which continues to push prices up, further increasing this gap. The economic reality for young people in the UK is very different to their parents, and the heuristics for long-term planning, and indeed future expectations, needs to change. Once we have a situation where a significant number of children are born to parents who don't own property, this could result in a polarisation of society, and a concentration of wealth in the house-owning fraction. Hard work, education and even a good career, may not be enough to power social mobility. Perhaps this is already happening.

One apparent axiom of financial planning appears to be that buying a house should always be what you should do if you are able. If you have a job that requires mobility and/or provides accommodation, then buying a house and renting it out is perhaps the best option of all. This seems fairly obvious, but is it?

Pros:

- Although the increasing unaffordability of houses in the UK may not always sustain house prices rises,[14] the analysis in the earlier part of this chapter implies it is possible to make money overall with a mortgage over the long term, and that is not even including rental income. Buying a house with the minimum deposit,

---

[13] Alas in a fictional Universe as of 2021!

[14] If the majority of housing stock is owned by increasingly older people, and the housing stock represents an increasingly larger slice of the financial resource, and more people appear to require expensive nursing homes in their later years, does this imply an eventual flood of house sales, which could reduce house prices in the next 10–20 years?

and charging enough rent to service the mortgage, is surely an obvious get-rich-not-so-quick scheme?

- Owning a house that is rented can be a handy insurance policy if one runs into health, domestic or career problems. You can cease renting and live in it, and if the house is large enough, rent rooms out to lodgers to provide a smaller, but not insignificant income.

Cons:

But we have not included the costs of owning property! The profits of an investment scheme will be subject to taxation, and also some form of professional management fee will need to be included. For housing, the costs might be:

- The costs involved in buying and selling (e.g. lawyers fees, estate agent fees, surveyors fees, stamp duty, etc.).
- Ongoing physical maintenance costs of the building, and may include utilities, ground-rent, shared-ownership management fees, etc. The home-owner may not be in a strong position to negotiate these costs, and therefore is at financial risk if significant rises are applied.
- Council tax (i.e. regional rather than national taxation. In the UK, contribution bands are semi-aligned to local house prices).
- Management fees of companies, if you use one to manage tenants, advertise the property, etc.
- A buy-to-let mortgage is often more expensive than a personal one.
- You will have to pay tax on the rental income.
- If you sell a buy-to-let property you will have to pay capital gains tax on the profits, subject to an allowance each year.
- And not to mention a very significant amount of time and worry! Do you want to spend your weekends and holidays looking after your rental property?
- Also, is it ethical to purchase a house that you don't live in, and as a result prevent someone of lower income than you from acquiring property? Buying a house to rent, and making a point of being a benevolent landlord, is surely socially better than holiday-home ownership, which can erode rural and coastal communities even more effectively than the wind and tide. However, the investment nature of a buy-to-let transaction (and the desire for profit) is perhaps yet another driver of house-price rises.

Note property investors *don't* tend to have fixed-rate repayment mortgages, but instead opt for the interest-only variant. These are typically going to be lower in interest rate than the equivalent repayment variety. The idea is the rent should more than compensate for the mortgage + associated costs, and any surplus can be reinvested in a savings scheme (or perhaps another property). This is all very well unless house prices fall, although if you don't sell the house this fiscally problematic scenario can be avoided. If house prices continue to rise, then you will gain if you sell the house as well as having benefitted from rent. The strategy here is clearly to minimise the initial deposit, i.e. the proportion of the house you actually own. Property investors are essentially paying for the right to charge rent on a house, and accepting the responsibility for the maintenance of the asset and the requirements of the mortgage.

11.5 A Simple Stock-Trading Model

I'm going to begin with a massive caveat. I know nothing about stock-picking, and I am far too nervous about losing all my savings to have a go myself. In my darker moments I am inclined to share the belief of Leonardo di Caprio's character in the *Wolf of Wall Street*, who begins his descent from honest broker to fraudster by buying into the cynical idea that the stock market is some form of collective delusion of wealth creation, that only benefits those that turn and oil this gaudily lit merry-go-round. However, if you have a bank account, run or work for a corporation, draw or contribute to a pension, and assume insurance will pay out, then you too are invested in the global financial system. I'm pretty sure the Jordan Belforts, and the perhaps nominatively deterministic Bernie Madoff's of this world are the small number of bad eggs in a high quality Marks & Spencer battery farm of financial workers. So how is it possible to make money from the trading of shares,[15] and how might you develop an algorithm to achieve this? Michael Lewis' *Flashboys* [25] describes the concept of *High Frequency Trading* (HFT), where computer programs execute thousands of trades every second, far-outstripping the speed that human traders can react to market information. My model, `zbroker`, will be much more pedestrian (i.e. trades are executed on a daily basis), but will be similar in a sense to HFT in that is is based upon a computer algorithm rather than human intervention. The idea is based on the naive assumption that assumes stock prices are Normally distributed within a moving window of D days. The 'bell-shaped' curve of a Normal Distribution is the limiting curve of many quantities subject to random variation. For a random variable x with mean μ and standard deviation σ, the probability of x being in range $x \to x + dx$ is $p(x|\mu, \sigma)dx$ where:

$$p(x|\mu, \sigma) = \frac{1}{\sqrt{2\pi\sigma^2}} e^{-\frac{(x-\mu)^2}{2\sigma^2}} \tag{11.21}$$

The vertical bar | means 'given', i.e. $p(x|\mu, \sigma)$ means the probability density of variable x given parameters μ and σ. The peak of the bell curve $p(x|\mu, \sigma)$ is at $x = \mu$ and the width of the curve scales with σ. To calculate normally-distributed

---

[15]Shares (e.g. the 'expectation of future value' of, in our examples, UK FTSE 100 companies) are of course but one of a plethora of investment vehicles. In addition to property there are bonds, 'futures', 'derivatives', gold and other precious metals, fine art, wine … Essentially anything that has *the potential to be worth more in the future*.

probabilities, we *standardise* the distribution using the substitution $z = \frac{x-\mu}{\sigma}$. In this case $p(z) = \frac{1}{\sqrt{2\pi}}e^{-\frac{1}{2}z^2}$. In zbroker:

$$z = \frac{\text{stock price} - \text{mean value of stock price}}{\text{standard deviation of stock price}} \tag{11.22}$$

and means and standard deviations (*volatility*) are computed from a moving average (using a fixed window size D in days) of past FTSE data. If something is normally distributed, the probability of the variable being greater than the magnitude of a certain z value is increasingly unlikely as the magnitude of z increases. In zbroker, the idea is that if a share price has a z value which belongs to the 'tail' of the normal distribution, then it is probably either undervalued (so the stock should be bought) or overvalued (so the stock should be sold). In the simulation illustrated in Fig. 11.7, I set z_{sell} and z_{buy} to be the same magnitude (but opposite signs), and vary this magnitude in unit steps from 1 to 4.

In the standard normal distribution, the probability that a z value is less than Z is given by:

$$P(z < Z) = \frac{1}{\sqrt{2\pi}} \int_{-\infty}^{Z} e^{-\frac{1}{2}z^2} dz = \tfrac{1}{2} + \tfrac{1}{2}\mathrm{erf}\left(\tfrac{1}{\sqrt{2}}Z\right) \tag{11.23}$$

where the *Error Function*, which is available as a numerical recipe[16] in most computer software, is:

$$\mathrm{erf}(x) = \tfrac{2}{\sqrt{\pi}} \int_{0}^{x} e^{-t^2} dt \tag{11.24}$$

Hence:

$$P(z > z_{\text{sell}}) = \tfrac{1}{2} - \tfrac{1}{2}\mathrm{erf}\left(\tfrac{1}{\sqrt{2}}z_{\text{sell}}\right) \tag{11.25}$$

$$P(z < z_{\text{buy}}) = \tfrac{1}{2} + \tfrac{1}{2}\mathrm{erf}\left(\tfrac{1}{\sqrt{2}}z_{\text{buy}}\right) \tag{11.26}$$

where z_{sell} is a positive value like 1, 2, 3, 4 and z_{buy} is a negative value like -1, -2, -3, -4. A graph of the standard normal distribution $p(z) = \frac{1}{\sqrt{2\pi}}e^{-\frac{1}{2}z^2}$, and plots of $P(z > z_{\text{sell}})$ vs z_{sell} and $P(z < z_{\text{buy}})$ vs z_{buy} are illustrated in Fig. 11.8.

The analysis performed, and core algorithm of zbroker, is as follows:

- Download from the internet[17] the closing stock prices (in pence, i.e. £0.01) of the FTSE 100 companies registered in the UK stock exchange in London over the

---

[16]The *Error Function* $\mathrm{erf}(x)$ can be approximately evaluated by integrating terms of the Maclaurin expansion of e^{-t^2}. This polynomial expansion can then be evaluated by a computer, to arbitrary precision depending on the number of terms used. $\mathrm{erf}(x) \approx \frac{2}{\sqrt{\pi}}\left(x - \frac{1}{3}x^3 + \frac{1}{10}x^5 - \frac{1}{42}x^7 + \frac{1}{216}x^9 - \cdots\right)$.

[17]https://uk.finance.yahoo.com/quote/%5EFTSE/components?p=%5EFTSE.

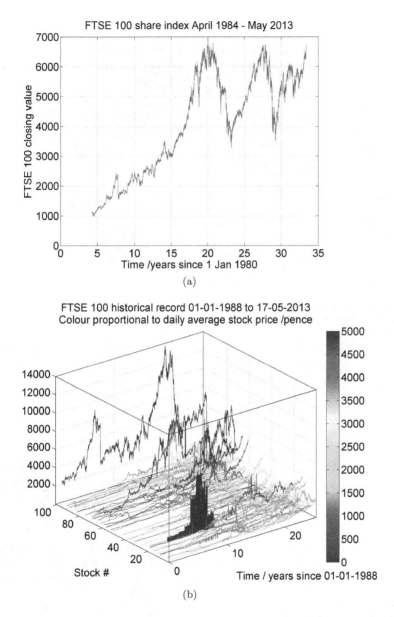

Fig. 11.5 (a) Closing values of the FTSE 100 share index of UK (public) companies between April 1984 and May 2013. The cycles of 'boom and bust' tempt certain soothsaying economists to believe that trends are predictable. I am not so sure! Graph (b) tracks the actual share price (in pence, i.e. £0.01) of the individual companies that comprise the FTSE 100. Significant volatility in price means it is possible to gain, or lose, significant sums simply by trading shares. *Source*: https://uk.finance.yahoo.com/quote/%5EFTSE/components?p=%5EFTSE.

period 1988 to 2019. The variation of stock prices, both in terms of the FTSE 100 all-share index and individual stocks, is illustrated in Figs. 11.5 and 11.6.

- Start with an initial cash fund size. In my simulations in Fig. 11.7 this is £1,000. Initially divide up the fund equally amongst all stocks, weighted by stock price.

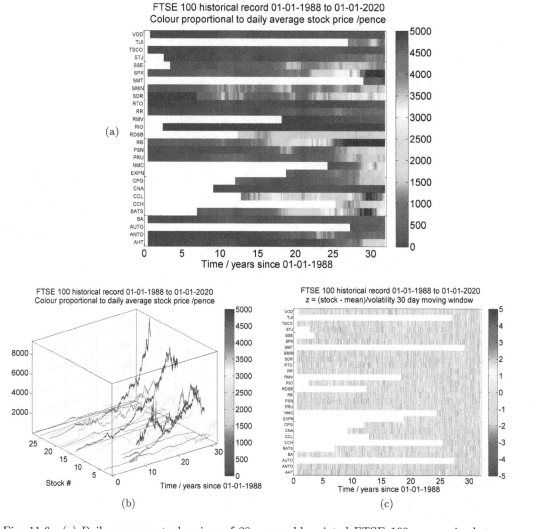

Fig. 11.6   (a) Daily average stock prices of 28 name-abbreviated FTSE 100 companies between January 1988 and January 2020. VOD means 'Vodaphone', BA means 'British Airways' etc. The colour scale indicates stock price in pence, and the white gaps mean a company was not in the FTSE 100. Graph (b) is a 3D version of (a). Graph (c) is a plot of the z value for each stock vs time in years since January 1988. z is the ratio of the difference in stock price from a 30-day moving average, divided by the standard deviation (i.e. 'volatility') of stock price in the same time frame. Significant variation in z from -5 to $+5$ suggests that prices often swing to the extremes, if one were to assume prices are random and normally distributed. A large $|z|$ shall be used as the basis of a stock-trading algorithm.

- Assume a transaction loss of L % of stock price. This is assumed to be the same for selling or buying. Assume transactions occur on the same day, i.e. there is no delay on a daily scale.
- Each day have a transaction fund of T % of the current stock portfolio. Each day sell maximum T percent of portfolio with z values above z_{sell}, and use the cash

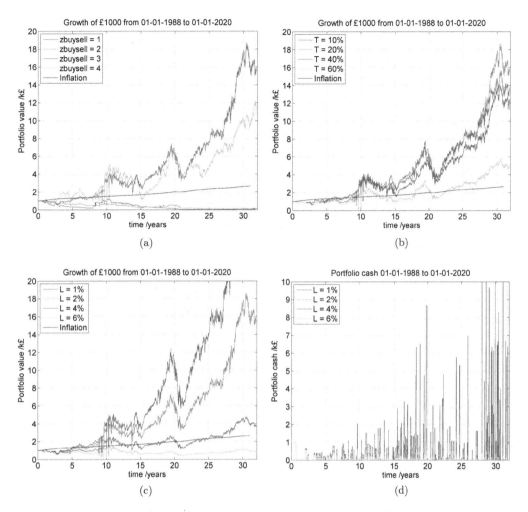

Fig. 11.7 Example of a stock-trading algorithm applied to historical FTSE 100 share prices. The idea is to fix a z value threshold, where z is the ratio of the difference in stock price from a moving average, to a stock price standard deviation calculated over the same time window as the moving average. If z exceeds the threshold we sell a stock, if z is below a threshold we buy a stock. Caveats are: (i) We have a transaction fund which represents T % of our portfolio value. Stocks that meet the upper z threshold criteria are sold and converted to cash each day. The cash is then used to buy stocks that satisfy the lower z threshold criteria. It is assumed every transaction carries a fixed loss percentage L due to fees, market volatility, etc. Graph (a) illustrates the effect of different z thresholds. $L = 2\%$ and $T = 20\%$. The effect of inflation is also plotted. Graph (b) compares different T values. z threshold is ± 3 (symmetrically for buy and sell) and $L = 2\%$. Graph (c) compares different L values. $T = 20\%$, z threshold $= \pm 3$. Graph (d) involves the same modelling as (c), but plots the amount of portfolio cash (i.e. the value of the transaction fund) over time.

from these sales (noting the transaction loss) to buy from the market those shares with z values below z_{buy}, up to the value of the cash fund.

- If a stock is removed from the FTSE 100 it is immediately sold. Its closing value will be added to the portfolio value.

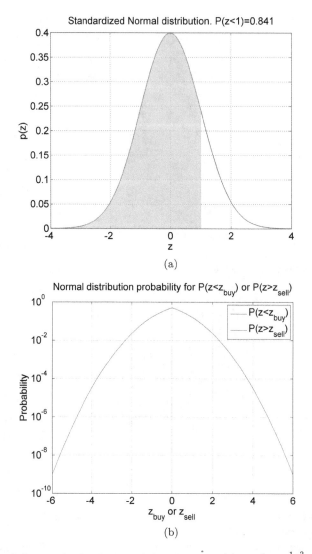

Fig. 11.8 (a) Plot of the standardised Normal distribution $p(z) = \frac{1}{\sqrt{2\pi}}e^{-\frac{1}{2}z^2}$ vs $z = \frac{x-\mu}{\sigma}$ where x is a Normal-distributed random variable with mean μ and standard deviation σ. The probability $P(z < 1) = \int_{-\infty}^{1} p(z)dz = 0.841$ is illustrated by the green area under the curve z vs $p(z)$. The total area under the curve must be unity, in order for $p(z)$ to represent a continuous probability distribution. (b) Plot of $P(z < z_{buy})$ vs z_{buy} (the negative values of z_{buy}) and $P(z > z_{sell})$ vs z_{sell} (the positive values of z_{sell}). The diminishing likelihood of $|z|$ being larger than increasing larger values of $|z_{sell,buy}|$ is the basis for a sell or buy rule employed by the **zbroker** algorithm.

- The model assumes the total volume of shares is much higher (and will be unaffected by) the buying and selling of stocks. In other words, it assumes **zbroker** is managed by a small amateur like myself and not Goldman Sachs, Warren Buffet or George Soros. One assumes when these players make large trades, the market moves too. This complex feedback mechanism would be much harder to model, although no doubt the major players invest heavily in 'quants' (e.g. physicists put out of work when a huge new particle accelerator project is cancelled[18]) to have a go. Even a tiny advantage may result in huge financial gains.

The outputs to **zbroker** are:

- A graph of portfolio valuation in £k vs time /years.
- A matrix of portfolio stocks held vs time /years. Matrix elements are (i) number of shares and (ii) total value.

zbroker conducts a 'what if?' analysis on historical stock data, although importantly involves no *a priori* model of stock performance (e.g. following a close financial study of a given listed company) aside from an assumption of normally-distributed stock price within a suitably small time window. It is a coldly numerical, high-ish frequency trading approach, and stocks are only kept if they are not deemed over or under-valued. No long-term clairvoyance is attempted. **zbroker** could easily be deployed with real stock data as an input, and need not be restricted to a daily timestep. If the Normal distribution assumption is deemed valid, the moving average window could be shrunk to D minutes, seconds, microseconds, etc.

The results from the simulation illustrated in Fig. 11.7 are rather tantalizing. The FTSE 100 data is illustrated in Fig. 11.6, a random selection of 28 stocks, tracked by daily closing prices between January 1988 till January 2020. $D = 30$ days is used to compute the moving averages and hence the z values. Subplot (a) of Fig. 11.7 illustrates the effect of different $z_{\text{sell,buy}}$ thresholds. The transaction loss is fixed at $L = 2\%$, and the transaction fund is fixed $T = 20\%$ of the total value of the portfolio. The effect of inflation is also plotted to give an idea of whether the investment strategy was successful. If the overall growth curve of the initial £1,000 investment isn't consistently beyond the inflation curve then the investment hasn't really added any additional financial value. Figure 11.7(b) compares different transaction fund T values. $z_{\text{sell,buy}}$ threshold is fixed at ± 3 (symmetrically for buy and sell) and $L = 2\%$. Figure 11.7(c) compares different L values, with $T = 20\%$, and $z_{\text{sell,buy}}$ threshold $= \pm 3$. Figure 11.7(d) involves the same modelling as Fig. 11.7(c), but indicates the amount of portfolio cash held.

Figure 11.7(a) suggests a higher value of $z_{\text{sell,buy}}$ results in a significant change in fund performance. $z_{\text{sell,buy}}$ values less than 2 cause the fund to lose value. The shares traded don't result in significant gains on average, and therefore the transaction loss eventually wears the fund down. However, a $z_{\text{sell,buy}}$ value of 3 yields a sixteen-fold

---

[18]There is a rather darkly humoured joke that one of the main causes of the global financial crash in 2008 was the cancellation in 1993 of the *Superconducting Super Collider* (SSC) particle physics project in the United States, ironically nicknamed 'The Destroyer', which was a potential rival to the *Large Hadron Collider* currently in operation at CERN in Switzerland. The thousands of physicists involved were then typically re-employed by investment banks. And every physicist just can't resist making ever more complex mathematical models. . .

growth! A forensic analysis of the shares traded has not been attempted (although **zbroker** does log all the trades so this could be done), and it is therefore not clear if this is mere luck due to a random bet on a stock that rises spectacularly, or a summation of many, but slightly less fortunate picks. Interestingly, a $z_{\text{sell,buy}}$ value of 4 results in a much lower portfolio performance than $z_{\text{sell,buy}} = 3$, although still a healthy 12-fold return rather than a loss. The inference here is that $z_{\text{sell,buy}}$ might have an *optimum* value. If **zbroker** were to be used with real stocks, perhaps historical scenarios could be run every $100D$ or so time intervals, and the best $z_{\text{sell,buy}}$ values chosen accordingly. In other words, **zbroker** could adapt to longer term market volatility conditions by changing its rules on a longer time scale than the window used to compute the moving averages and hence the z values.

Figure 11.7(b) indicates an optimum value of T might also exist for our dataset. Raising T to about 20% appears to result in the best returns, and if a larger value is set, the portfolio performance reduces. Indeed if $T = 60\%$ is set as the maximum fraction of portfolio value to be transacted in a day, the performance is only a little better than inflation, and sometimes it is worse.

Figure 11.7(c) and 11.7(d) indicate fairly obvious trends. Unsurprisingly, as the transaction loss % L reduces, the portfolio gains significantly. The key point is that for a portfolio gain to be better than inflation, L needs to be less than about 4%, given 'crudely optimal' parameters of $T = 20\%$, and $z_{\text{sell,buy}}$ threshold $= \pm 3$.

Chapter 12

Power to the People

If you are reading this from outside the United Kingdom, and are confused by the strange British pop-cultural references that I have managed to sneak past the publishers, enjoy a little bit of *schadenfreude* at the expense of blustering Blighty, resulting from the bizarre political paroxysms that she has inflicted upon herself in the past decade. General Elections in the UK usually take place every 5 years. However, an ongoing soap opera of leadership contests, and a worrying left and right polarisation of both main parties (Conservative and Labour) from the dull but reasonable centreground, all fuelled by a very close 52% to 48% decision by referendum in 2016 to leave the European Union, have resulted in *four* elections: 2010, 2015, 2017 and 2019. I do not propose to add any more words to the frothing torrent of impassioned opinions that have been our national narrative for the past decade. (Well, prior to the coronavirus pandemic in early 2020, which understandably has eclipsed all other discourse). My aim in this chapter is instead to develop some simple tools to make sense of the statistics and infographics that stare out of newspapers and print magazines and the perennial pictograms that clamour for attention on our computer screens, televisions and smartphones. I also want to explore the idea of fairness in UK General Elections, in the sense of how the political affiliations of our 650 Members of Parliament (MPs) reflect the actual proportions of the vote. I will conclude by entering into some speculation how a *proportional* system might be constructed in practice, and in doing do, show that there might be some uncomfortable unintended consequences. If I do entertain a semi-political opinion myself, it is that one should endeavour to understand reality via the data, and the details of the consequences of modelling, before making an ideological stance. Or better still, reject ideology entirely and simply aspire to be a realist and a pragmatist. This might make you more respectful and empathetic towards those with differing opinions to your own, or may fuel your ire even further. I'll leave that up to you to decide.

12.1 A Very Short Guide to UK General Elections and Parliament

In order to make sense of the analysis of UK General Elections 2010–2019 let us summarise the key features of UK democracy[1] germane to our discussion.

- During a General Election, the UK public elects 650 MPs to represent them in the *House of Commons*. A single MP is returned from each *constituency*. This is a geographic entity, involving a population of around 100,000 people. As shown in Fig. 12.3, the average constituency vote (based upon 2010–2019 elections) is between 45,500 and 49,400, i.e. around half the estimated total population. This roughly tallies with total (2019) UK voting turnout of about 32 million out of a total population of 66.7 million. (*Source*: https://www.ons.gov.uk/ accessed 15/4/2020). Regions with a significant proportion of urban areas such as London or South-East England will contain more constituencies. London has 73, whereas the whole of Scotland returns just 59 MPs, and Wales provides only 40. However, since 1997 there has been a devolved Parliament (or 'National Assembly') in Scotland and Wales, and since 1998 in Northern Ireland. These have significant, although not comprehensive, law-making and fiscal powers. At the time of writing in April 2020, the UK remains a single national entity. However, active campaigns persist, particularly in Scotland, to fully separate from the UK.[2] The fallout from 'Brexit' may raise the issue of Scottish independence to greater prominence in the next few years.

- The MP for each constituency is the candidate who wins the most votes during a parliamentary election. This is called the 'First-Past-The-Post' (FPTP) system. It can lead to quite significant disparities between the number of seats in Parliament and the proportion of the total vote for a given party. The lack of proportional representation is typically cited as a challenge to democracy in the UK, and to the mandate of an elected government. This can be particularly problematic given the leaders of political parties are typically elected by members of their own party, which can mean a Prime Minister can end up being chosen by a *very* small proportion of the population.[3] However, the FPTP system does help prevent representation from minority parties that might give legitimacy to more extreme views,[4] and also makes it less likely that Governments are formed from

---

[1] I summarise from: https://www.parliament.uk/ (accessed 15/4/2020).

[2] In a political sense! I don't think a vast dividing moat or extension to Hadrian's Wall is proposed, even from the most zealous members of the SNP.

[3] According to https://www.theguardian.com/politics/2019/jul/23/boris-johnson-elected-new-tory-leader-prime-minister, Boris Johnson beat Jeremy Hunt in the July 2019 leadership election by 92,153 votes to 46,656. This corresponds to $\frac{92,153}{66.7 \times 10^6} \approx 0.14\%$ of the UK population!

[4] As illustrated in Fig. 12.1, the UK Independence Party (UKIP), who campaigned strongly to leave the EU and, it is probably fair to say, are somewhat to the right of the political spectrum, gained 13% of the vote in 2015. The British National Party (BNP), a "far-right fascist organization"

fragile coalitions between parties with somewhat differing agendas. The latter could potentially be problematic for efficient decision-making in times of crisis. In other words, there are many pros and many cons to this system, and indeed *any* political system.

- The 650 constituencies are grouped into 12 regions. These are: *Yorkshire and The Humber, West Midlands, Wales, South West, South East, Scotland, Northern Ireland, North West, North East, London, East Midlands* and *East*. The number of votes cast per region (see Fig. 12.4) varies from less than one million (Northern Ireland) to over four million (South East).

- Not all UK (or qualifying Commonwealth) citizens can vote in General Elections. The voting age (as of April 2020) is 18 years, you need to register to vote in advance (which normally requires a valid UK address) and you must not belong to a group that is 'legally excluded' from participating. These include: members of the *House of Lords*,[5] convicted prisoners, anyone found guilty of corruption charges in the past five years, and those with a mental illness or disability deemed to prevent them from making a reasoned judgement. By *convention*,[6] the Monarch is politically independent, does not vote (or stand for election) as this would break constitutional norms. However, it is not clear to me whether there is actually a law against it, and how far the precedent of apoliticism penetrates the royal household, and what the consequences of this might be. I'm sure this issue will be addressed in Netflix's *The Crown* at some point![7] In the EU referendum in June 2016,[8] about 28% of the total population were not registered to vote, and a further 20% chose not to vote, even though they were registered. An estimated voting population of about half the citizenship is therefore expected, which tallies with the figures in the previous paragraph. The overall votes and percentages for

---

according to Wikipedia, gained 1.9% of the vote in 2010. However, UKIP gained just a single seat in the 2015 election, and BNP had zero MPs in 2010. Had full proportional representation been used, UKIP would have gained 83 seats in 2015, and the BNP would have returned 13 MPs in 2010 (see Fig. 12.2 for details).

[5] The *House of Lords* comprises about 800 individuals, and the majority are 'life peers' meaning that they do not need to seek re-election. The purpose of the House of Lords is to offer independent scrutiny to proposed new laws ('Bills'), although ultimately the Lords don't have the power to block them, particularly those involving financial decisions. Although both Lord (and Royal) ascent is technically required for a Bill to become Law, the will of Parliament is usually carried, albeit with agreed amendments.

[6] https://www.royal.uk/queen-and-government (accessed 16/4/2020).

[7] The novel *To Play the King* by Michael Dobbs considers a fictional future possibility of a Monarch who definitely *isn't* apolitical.

[8] https://www.indy100.com/article/brexit-leave-remain-52-48-per-cent-voter-turnout-electoral-regi ster-charts-7399226 (accessed 15/4/2020).

the 2016 EU referendum in the UK are provided in the table below:

		% of total	% voting	% of vote
Voted leave	17,410,742	27%	37%	52%
Voted remain	16,141,241	25%	35%	48%
Did not vote	12,948,018	20%	28%	—
Not on electoral register	18,099,999	28%	—	—
TOTAL	**64,600,000**	**100%**	**100%**	**100%**

- MPs are typically affiliated with a political party, with the largest (in 2020) being the Conservatives, Labour, Liberal Democrats and Green; Scottish Nationalist Party (Scotland only); Plaid Cymru (Wales only) and Democratic Unionist Party (DUP), Sinn Féin, Alliance and SDLP (Northern Ireland only). The party with the majority number of MPs (i.e. 'seats') following a General Election will form a Government, and the runner up becomes *Her Majesty's Opposition*. This is true in a very literal sense, as Government and Opposition tiered benches face each other across a small divide in both the House of Commons and the House of Lords. A number of MPs will be elevated to the role of *Minister* within a *Cabinet*, which is chosen by the *Prime Minister* (i.e. the MP who has been chosen as the leader of their party) and acts as a form of executive, although all laws must be passed by Parliament.[9] There are twenty-two members of the (April 2020) Cabinet, comprising 21 MPs plus the Leader of the House of Lords.

- A *coalition government* is also possible if a single party does not achieve an outright majority following a General Election. This occurred in the UK between 2010 and 2015, when the Liberal Democrats formed a coalition with the Conservatives. However, the balance of power between coalition partners may not at all be equal. As shown in Fig. 12.2(a), the Liberal Democrats won just 57 seats in the 2010 election compared to 306 for the Conservatives, and 258 for Labour (who became the Opposition).

12.2 An Analysis of UK General Elections 2010–2019

12.2.1 *Data sources, and raw data vs information*

2010, 2015, 2017 General Election results (in spreadsheet form) were downloaded[10] in 2019. Unfortunately this link appears to have expired as of 04/2020, and a search

---

[9] As of 2009, there is also the *Supreme Court* of the United Kingdom, which controversially overruled Prime Minister Boris Johnson's attempt to *prorogue* (i.e. suspend) Parliament for five weeks prior to the State Opening of Parliament on 14th October 2019, which would have significantly reduced the amount of time to debate the Government's plans regarding 'Brexit', i.e. the departure from the European Union. 'Brexit' was eventually achieved on the 31st January 2020.

[10] https://data.gov.uk/dataset/b77fcedb-4792-4de4-935f-4f344ed4c2c6/general-election-results-2017.

for General Election in https://data.gov.uk brings up a long list of disparate sources rather than a single consolidated source. This 'fragmentation' problem chimes with the comments of Rogers (2013) in *Facts are Sacred* [40]. Data may have been published in various forms, but until it is assembled into a spreadsheet or similar it is very difficult to analyse, and the gap between data (just the facts) and information (what the facts might mean) is widened. Internet-based data can be particularly problematic as links may change or expire without warning, and extra protocols must be followed to prevent malicious or accidental modification. The problem regarding the publication of raw data is even wider than my brief encounters with election results. Even in peer-reviewed scientific papers, the published data is often highly processed, and might just be a line on a graph or a single figure in a table. It is not always so easy to access the raw data. A key feature of science is that all experiments should be repeatable, and that goes for the analysis too. I hope in a small way to remedy this problem by publishing *all* the raw data and associated spreadsheets and computer code for generating the figures in this book on my website: www.eclecticon.info/ scibysim. If you decide to construct the graphs yourself using your own software, then the core purpose of this book will have been achieved.

The 2019 election data source was a spreadsheet,[11] and this link appears to be live as of 04/2020. However, it is worth mentioning a problem I have encountered in every single analysis task I have performed over the years. The formatting of the data is *never* consistent. I will spare you the details outlined in the Methodology subsection below, but I always find I have to write bespoke code *every time*. The format of the 2019 election data spreadsheet was significantly different from the 2010–2017 data, and as a result I had to write a distinct 'ingester.' This can sometimes be more onerous than the analysis which follows. In my case this was mildly inconvenient, but in general I imagine it creates an additional barrier to freedom of information, and perhaps even to democracy itself. It is much easier to combat 'fake news' if access to official data sources is both open *and* the format of the data is such that it facilitates independent analysis. In other words, raw data in spreadsheets + processed data in electronic documents, with a detailed recipe of how the processing was done so it could be repeated.

12.2.2 *Methodology*

I like to use the software MATLAB to perform processing and data visualising tasks. Nearly all the graphs and simulations in this book are constructed using this tool. The benefits of MATLAB are that it combines a programming language (i.e. a series of precise commands typed into a text file that can be executed, rather than the constraints of a graphical user interface), it contains a huge library of

---

[11] https://researchbriefings.parliament.uk/ResearchBriefing/Summary/CBP-8749.

built-in functions that offer mathematical and graph-making tools, and has a large international community of users in Engineering and Science fields. However, the actual software platform itself doesn't really matter,[12] as long as the tool you wield works effectively and fits with your particular tastes and prior-training. It is the *process*, the implementation of a *data-processing-pipeline*, which is important.

To analyse the 2010–2019 UK General Election data I implemented the following two-step processes: (1) Data ingest: Excel spreadsheets to MATLAB arrays, then (2) Data analysis and automated infographic production.[13]

12.2.3 *Data ingest: Excel spreadsheets to MATLAB arrays*

Ingest of the election data from two spreadsheets into an *array* format. In essence, these are 'internal' spreadsheets that can be manipulated using *code*. `parties` is a 227 × 1 *cell* array[14] of the set of 227 unique party names who were listed in the source spreadsheets. Only a small number of these gained seats, and I have used a 0.2% cutoff in several of my graphs in order to focus upon the main voting preferences, i.e. the 'signal, not the noise', as Nate Silver might put it [44]. `constituencies` is a similar cell array of the names of all the UK constituencies. Interestingly, 651 were listed rather than 650. For each election year: `total_vote_by_constituency` is a 651 × 227 array of actual votes, with the rows corresponding to the constituencies, and the columns to the parties. A similar `total_vote_by_region` of dimensions 12 × 651 records the total votes summed over geographic regions, which are listed in a 12 element cell array `regions`. From these arrays, all other statistics can be computed by computing averages over sums or columns. To assist in analysis, I also created additional *structured arrays*[15] for each region and each constituency. `regions(4).votes_by_constituency` is, for example, a 29 × 227 array of votes for the 29 constituencies in the

---

[12]You could use *Python, Octave, Javascript, C++, Perl, R*, or a plethora of other systems.

[13]Separating data ingest from analysis and infographics production is a sensible idea if the ingest stage is not essentially instantaneous. You will discover that you will need to run the graphics production software *many* times before you find all the bugs in the code. A lengthly, repeated, ingest stage will make this process tedious. Once ingested into data arrays, data manipulation in software such as MATLAB can be impressively rapid.

[14]In MATLAB terminology, arrays can be numeric, character or cell forms. Numeric [...] arrays are vectors, matrices or higher-dimensional data structures with elements that are all numbers. MATLAB copes with complex numbers natively, which is not true of all languages. Character arrays correspond to text characters like 'a' or 'z'. They actually correspond to the ASCII integers codes, and every time we use a character on a computer screen we actually process a number. Cell arrays {...} can have mixed elements, which can include other arrays too.

[15]A structured array is a fantastic method of keeping track of data when individual cell arrays could easily become unwieldy. `regions(n)` returns the nth element of a data structure `regions`. Two example fields are `regions(n).region`, which gives the text string of the region (e.g. 'London'), and `regions(n).vote_percent`, a vector of vote percentages for each of the listed parties.

'North East'. `region.constituencies(628).constituency` = 'Winchester', and `constituencies(628).votes` is a 1×227 array of votes for all 227 named parties. All the ingested arrays are then saved into a data file (`.mat`) format which can be loaded into MATLAB for the purpose of generating plots much faster than trying to deal with a giant spreadsheet in Excel or equivalent.

12.2.4 *Data analysis and automated infographic production*

We shall apply a data analysis process, in a loop, for all four election years. For each year a named folder is created, and 300 dpi PNG prints of the various analysis graphs are automatically written to this folder. The figures below are assembled manually from these images using a graphics package.[16]

(a) Horizontally stacked bar graph of the total vote %

Figure 12.1 is a horizontally stacked bar graph of the total vote %, in descending order, and subject to a cutoff of 0.2%. The bar labels are automatically generated from a cell array of party abbreviations, and comprise the subset of parties which exceed the total vote cutoff. Bar colours are set based upon exact RGB values, (in range 0 to 1,) returned by a function with input being the party name string, e.g. 'Labour' red is `d.L_RGB` = `[146,0,13]/255`. In the title of each graph is the year and the total number of votes cast.

(b) Horizontally stacked bar graph of parliamentary seats

Figure 12.2 is similar to Fig. 12.1 but instead plots the parliamentary seats along the horizontal axis. For each party there are three bars. The first is the actual number of seats won via First-Past-The-Post. The second bar (with the 'PR') label denotes the seats that would have been won if Proportional Representation was applied, i.e. MPs from each party determined on a national basis following the overall vote. It is not quite true PR, since one cannot have a non-integer number of MPs. It is better to be well rounded than a dis-member of parliament (!) The third bar (with the 'PRextra' label) relates to some personal speculation, detailed in the following section. In essence, how many *extra* MPs we would need to achieve a near-PR distribution, while also always returning the winner of each constituency vote. The extra seats are added to the FPTP actual election result, resulting in approximately the same distribution as PR. The immediate problem with this approach is that it enlarges Parliament considerably, perhaps doubling current numbers. However the general purpose is served: how can we illustrate, as clearly as possible, the impact of a policy change, and compare it to the status quo.

---

[16]I use *Xara Designer Pro*.

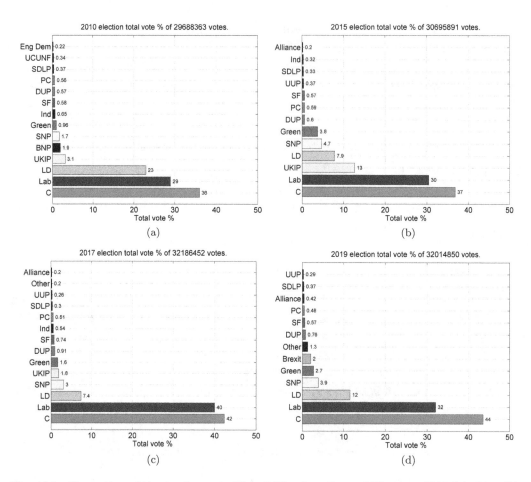

Fig. 12.1　Proportion of the total vote in United Kingdom General Elections 2010 (a), 2015 (b), 2017 (c) and 2019 (d). Letter code represents the political party, e.g. C means 'Conservative', Lab means 'Labour', LD means 'Liberal Democrats,' etc. *Source*: 2010, 2015, 2017: https://data.gov. uk/dataset/b77fcedb-4792-4de4-935f-4f344ed4c2c6/general-election-results-2017.

(c)　Vertically stacked bar graph of votes per constituency

Figure 12.3 is a vertically stacked bar graph and gives an idea of the total votes by constituency, and some idea of their variation (i) in total and (ii) between the parties. As discussed above, the key takeaway here is the idea of how many people actually voted, relative to the overall population, and hence an idea of the sizes of a constituency on average. Using the 2016 EU referendum as a guide (see above), if 20% were registered but didn't vote, 28% weren't eligible, then this leaves 52% that did vote. If the average constituency size if x, then $0.52x = 49{,}200$ (using 2019 election data), i.e. $x = 94{,}600$ which is fairly close to the 100,000 per constituency estimate. The table below was calculated from the 2019 constituency data. V is mean average votes per constituency (in thousands) per party, and σ_V is the standard deviation. The relative size of σ_V to V is quite telling in its size. The anomalous

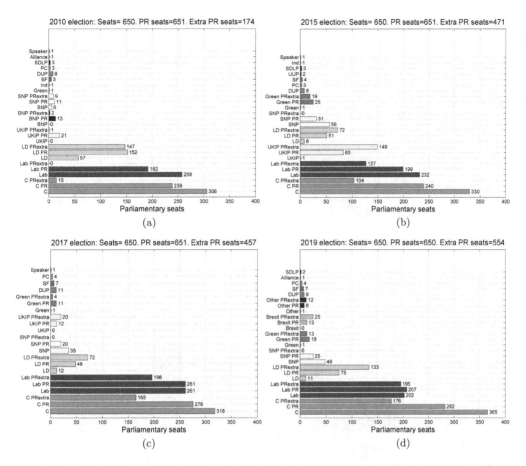

Fig. 12.2 Comparison between actual UK parliamentary seats gained (the lower number) in a General Election, the number of seats that would have been won given Proportional Representation (PR), and 'Extra' seats added to get the ratios of seats closer to the true PR fractions. This is needed because MPs are always discrete, well at least arithmetically. Note the extra seats are added to the FPTP actual election result (i.e. *not* the PR number) resulting in approximately the same distribution as PR. So for Conservatives in 2019, if we include extra PR seats, the proportion of MPs should be $(176 + 365)/(650 + 554) = 45\%$, which is close to the (rounded) PR proportion of $282/650 = 43\%$. The true PR proportion, based upon the national vote, should be 44%.

ratio is the SNP, which can easily be explained by only Scottish constituencies voting for the SNP.

Party	V	σ_V	$\frac{\sigma_V}{V}$
Conservative	21.5	10.3	0.48
Labour	15.8	8.8	0.57
SNP	1.91	6.2	3.25
Lib Dem	5.68	5.8	1.02
Green	1.33	1.8	1.35
Brexit	0.99	1.7	1.72

(12.1)

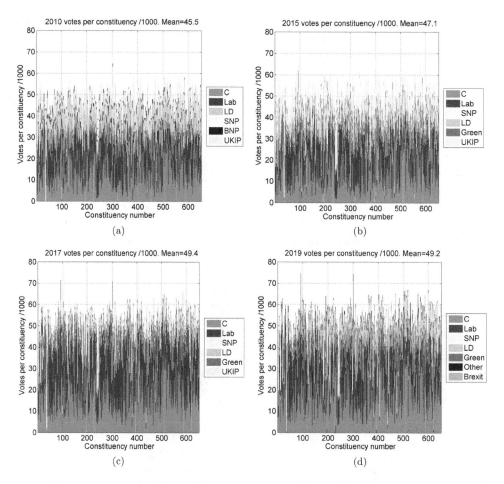

Fig. 12.3   Stacked bar graph of votes (in thousands) for each of the 650 constituencies in the UK, with each returning a single MP of the (colour coded) political parties. The chosen MP, as per the 'First-Past-The Post' system, is the MP with the most votes in each constituency. Constituencies are semi-geographical areas containing approximately 100,000 people. This tallies with the voter returns; in 2019 nearly 50,000 voted on average per constituency, and 32 million people voted out of a total population (including those ineligible to vote) of about 66.7 million. (*Source*: https:// www.ons.gov.uk/ acceded 15/4/2020). Large cities with high population densities such as London or Birmingham will comprise many constituencies. London has 73, whereas large geographic areas (but less dense in population terms) like Scotland return 59 MPs, and Wales only 40.

(d) Horizontally stacked bar graph of regional votes

Figure 12.4 is a horizontally stacked bar graph and represents the votes per party, in millions, per region. There are only twelve regions compared to 650 constituencies, so it is much easier to make out the individual party votes compared to the plot in Fig. 12.3. The jagged black line tracks the totals per region. The table below was calculated from the 2019 constituency data. V is mean average votes per region (in millions) per party, and σ_V is the standard deviation. The relative size of σ_V to V is again quite telling in its size, indicating significant regional variation. The anomalous

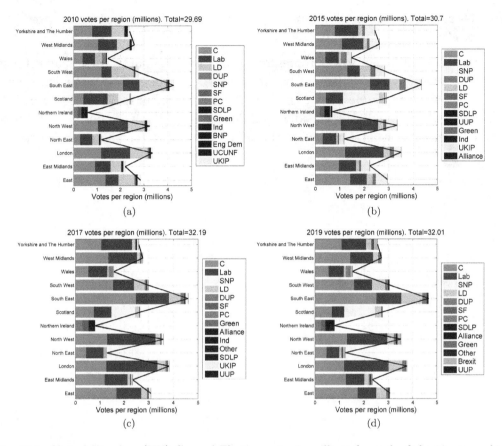

Fig. 12.4 United Kingdom (UK) General Election votes in millions for each of the 12 geographic regions in the UK. The horizontally stacked bar graphs correspond to the (colour coded) political parties who gained more than 0.2% of the total vote.

ratio is the SNP, which can easily be explained by only Scottish constituencies voting for the SNP. The actual numbers in V need to be interpreted carefully though. Unlike the constituencies, which are assumed to be of similar population of about 100,000, this is certainly not true of the regions. The 'South East' might have about four times the voting population as 'Northern Ireland'. This is a good example of when the graph conveys much less potential bias than processed data, such as in the table below.

Party	V	σ_V
Conservative	1.16	0.64
Labour	0.856	0.47
SNP	0.104	0.34
Lib Dem	0.308	0.23
Green	0.072	0.05
Brexit	0.054	0.05

(12.2)

(e) Geographic 'blob diagram' representation of regional votes per party

Figure 12.5 is a different way of representing the total regional votes per party. I have used the 'blob' idea made famous by Hans Rosling's animated 'Gapminder' graphs in the excellent *Factfulness* [42], and also the annual infographic produced by *The Guardian* which details UK spending per department [40]. The concept is the *area* of the filled circles is proportional to the actual number of votes received per region,

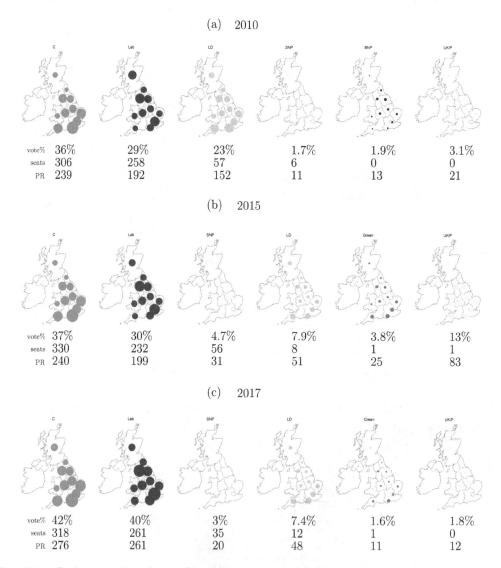

Fig. 12.5　Circles are plotted to indicate the total number of votes by region for the six most popular political parties in the 2010, 2015 and 2017 UK General Elections. The area of the circle is proportional to the number of votes.

per party. If we were to use radii instead, the circles would look disproportionately large or small. I have placed each circle in the centre of each geographic region to give a better sense of context, and have paired each national party map with the high-level statistics, i.e. vote %, number of seats, and the number of seats if proportional representation was applied. On visual inspection, the difference between Labour and Conservative appears quite marginal. The most obvious trends are the diminution of Liberal Democrat votes from their 2010 coalition success, and the rise and fall of UKIP.

(f) Geographic 'blob diagram' with constituency vote polygons

Figure 12.6 is essentially the same as Fig. 12.5, but instead of circles, each party is represented by a polygon whose 'vertex radii' are proportional to the number of votes per constituency. Particularly spiky polygons indicate significant variation in voting preferences within each region.

(g) All constituency vote polygons

Figure 12.7 uses the same polygon of constituency votes as in Fig. 12.6, but this time the regional component is removed, and all constituencies are represented, i.e. the polygon has 650 vertices. The black circle around the Conservative (blue) polygon – the winners in each of the four elections – is for scale, and represents 48,000 votes. On a national scale, all polygons are 'spiky', although perhaps the most interesting features are for the smaller parties. You can clearly see the 59 Scottish constituencies for the SNP (yellow), and no other votes elsewhere. You can also see a singular spike for the Green party, which corresponds to their only MP, Caroline Lucas, of Brighton Pavillion. There are also a few distinct spikes for the 2019 election 'Other' plot. For this dataset, all parties outside the main ones (e.g. independents, smaller special interest parties, etc.) were packaged into this category. 'Others' gained a single seat in the 2019 election, as indicated in Fig. 12.2.

(h) Geographic 'blob diagram' for the 2019 General Election

Figure 12.8 is the final graph and is particular to the 2019 election. Figure 12.8(a) is the coloured circle plot as per Fig. 12.5, representing by circle area, the number of votes per party by region. Total vote %, number of seats and 'seats if proportional representation were applied' is listed for each party. Figure 12.8(b) continues with the concept, but for the next six parties with the most seats. Figure 12.8(c) represents the regional constituency vote polygons for the top-six parties.

12.2.5 *Could FPTP be replaced by PR?*

An idealist might look at Fig. 12.2 and highlight some huge discrepancies between the overall vote and the seats in parliament achieved via First-Past-The-Post.

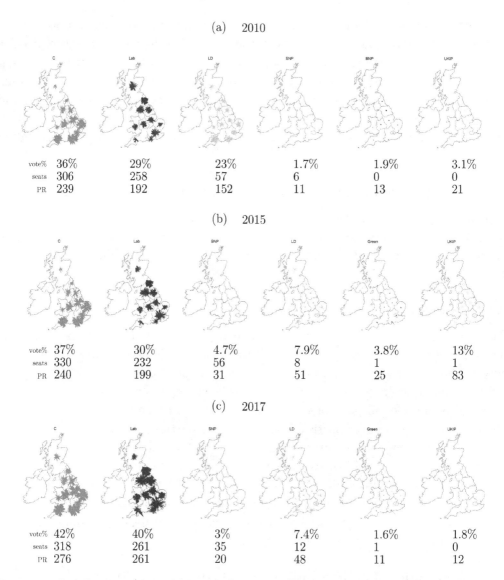

Fig. 12.6   Distribution of votes by region for the six most popular parties in the UK General Election 2010, 2015, 2017. Each region has an irregular polygon whose vertices correspond to the votes in each regional constituency. The radius of the vertex is proportional to the number of constituency votes. A spiky polygon corresponds to a region with very different voting preferences within its constituencies.

Taking 2019 as an example, the table below shows the ratio between actual seats and the PR predictions. Note I have omitted the Northern Ireland and Welsh parties, as in my PR calculations I have set a rule to exclude parties which get less than 1% of the total vote. In the 'PR speculation' below, the MPs returned for these parties would be unchanged from FPTP.

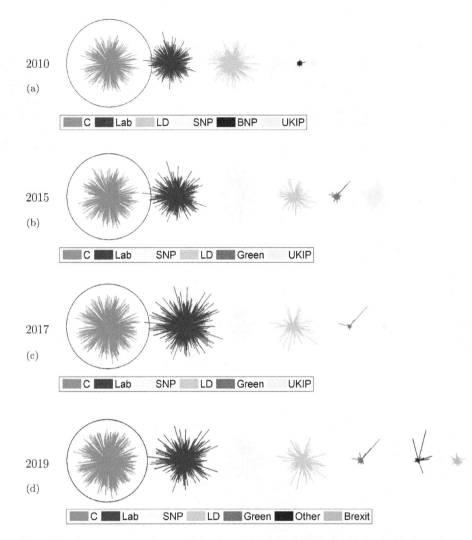

Fig. 12.7 Constituency votes per party in the UK General Elections 2010–2019 are represented by the spoke radii in these plots.

Party	Seats	PR seats	$\frac{\text{seats}}{\text{PR seats}}$
Conservative	365	282	129%
Labour	202	207	98%
Lib Dem	11	75	15%
SNP	48	25	192%
Green	1	18	6%
Brexit	0	13	0%
Other	1	8	13%
TOTAL	628	628	100%

(12.3)

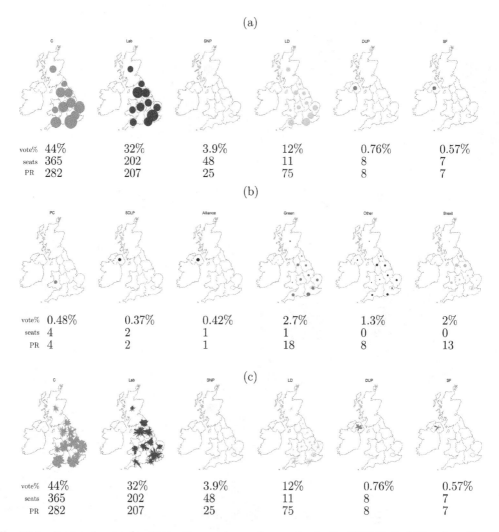

Fig. 12.8   Regional votes for the twelve most popular parties in the 2019 United Kingdom General Election are plotted in (a) and (b). The total vote percentage, number of seats, and proportional representation equivalent seats are also provided. (c) breaks down the regional votes into a polygon whose vertices represent the regional constituency votes. A particularly 'spiky' polygon indicates a region with a greater range of parties returned to parliament via the electorate.

In crude terms, the Conservatives gained 29% too many seats, and the SNP gained twice the number their proportion of the total vote should warrant. Labour is about right. (Although they are typically left!) While using percentages doesn't reflect the relative sizes of the votes for the smaller parties, they do illustrate a significant underrepresentation. Liberal Democrats, Greens and the Brexit Party all achieve 15% or less of the seats that a proportional system would suggest.

So could anything be done to remedy this situation? The obvious answer is to look outside the UK, and study the panoply of different electoral systems and weigh up their pros and cons in a similar way as I have done in the previous section. If you are reading this and feel inspired to embark on such a project, then please do so. There are probably whole libraries devoted to analysis of elections. So why continue with my speculation below? The point of this book is *not* to provide the final answer to a particular situation, i.e. the most refined model. Instead, my aim is to rigorously explore what is perhaps the *first* model that you might sensibly imagine. This then becomes the schematic that you can use to ask searching questions of more advanced models. So what would be a sensible first step to improve upon FPTP in the UK? Could we model that *in detail*, i.e. at the level of the actual MPs that would be selected in a PR election? In considering this, I started to think about the local MP, and in particular, their constituency rather than national party responsibilities. In the UK, power is devolved to a certain extent to local councils. My own is Hampshire County Council. Councillors are elected separately, and tend also to be affiliated to the same parties (Conservatives, Labour, etc.) as in national Government. Is the MP the head of the county council? No. Does the MP spend most of their time locally? No, they are predominantly in London, obviously. One could therefore argue that local MPs are merely sentient coloured chips in the weighted calculus of Parliamentary voting. Perhaps it doesn't matter who they are, and where they are from. Perhaps we should simply vote for parties, and then MPs are chosen (perhaps regionally) in the correct PR ratios from a list of suitably vetted candidates. The issue is that I *don't* believe there is a local disconnect, and that local connection *is* important. There must inevitably be a greater distancing for senior members of the Cabinet (e.g. Prime Minister, Chancellor, etc.), but I think the majority of MPs endeavour to represent their local interests just as much as they advocate the views of their party. Indeed, the concept of *local accountability* is probably vital to maintain the overall *reasonableness* of the political system in the UK. I imagine many scoffing loudly at this bold statement, but consider this: I think there would be much greater scope for corruption and alienation of large sections of society if your MP could *not* be directly voted out of office by fellow citizens numbering around 50,000. This is small enough to an MP to *know* a fair proportion of individual electors directly, albeit by one or two degrees of separation.

My thesis is that any change to FPTP would *have to maintain a local connection*. In other words, preserve the idea of constituencies of approximately equal population, and return the most popular MPs. Now we have a problem. For PR to work in 2019, 83 Conservatives who won their constituency vote would *not* be selected. If you start by filling up seats for the most popular party, you'll get to a situation where you need to find extra seats for a minor party that may have come third, fourth, fifth, etc. in a constituency vote. In other words, PR and local accountability are pretty incompatible. There might be a way round this by combining several nearby constituencies, so the impact of three MPs being returned who came first, second and fifth may not be so unpalatable. However, I think this would be a hard sell to the electorate. An alternative, which I will explore, is to *keep* the FPTP winners, and then select *extra* MPs to make up the true PR proportions. This is what the third 'PRextra' bar in Fig. 12.2 represents. I list some immediate pros and cons. There are surely many more.

Pros:

- You get even more local accountability than before. Perhaps the actual winner could be granted extra powers of local representation (e.g. they are automatically the chairperson for local matters if more than one MP per constituency is returned), to offset the concern of an MP with a low proportion of constituency votes being returned in order to make up national PR numbers.
- There are more MPs in Parliament, and subject to rounding errors, the number of MPs per party better reflects national political tastes. More MPs may mean greater diversity in opinion, and greater knowledge during debates.
- The Palace of Westminster won't be large enough for all the MPs. This could mean a number of regional chambers, connected via video link. This might help to burst the 'Westminster bubble', the perception that debate and decision making at a national level is far too London-centric.

Cons:

- Parliament probably doubles. MPs physically won't fit in the current House of Commons. The expanse of Parliament will need to double, the expense possibly by a greater multiple.
- There will inevitably be a greater representation of special interest and 'extreme' views. There will be a greater chance of coalitions which may enmire efficient decision making, and a possibility of an extreme party ending up holding the balance of power, despite having only a tiny electoral mandate. The threat of the latter is very real in modern day Europe. It is harder to imagine the possibility of a threat of this character, given that hasn't occured (in the UK) in living memory. Therefore this could be an *unintended consequence* of a well meaning but ultimately naive change.

I have modified the MATLAB analysis code used to create the figures in this chapter to also produce a text file (see Fig. 12.9) which lists every constituency, the FPTP MP, plus an allocation of the Extras. In addition to the high level statistics in Fig. 12.2, I discovered plenty of devil in the detail that might make this PR scenario somewhat unpalatable. A few unintended consequences were immediately rejected: e.g. more than one extra MP per constituency. In one model run, a Northern Ireland constituency ended up having five MPs! I also forced MPs who came worse than third to be rejected from the list. The trade-off is a slightly poorer fit to ideal PR proportions.

My algorithm for allocating extra seats is as follows:

(1) Select the subset of parties which achieved 1% or higher of the vote.

```
 1  Aberavon:                    FPTP MP=Lab, Extra MP=C (2)
 2  Aberconwy:                   FPTP MP=C, Extra MP=Lab (2)
 3  Aberdeen North:              FPTP MP=SNP, Extra MP=C (2)
 4  Aberdeen South:              FPTP MP=SNP, Extra MP=C (2)
 5  Airdrie and Shotts:          FPTP MP=SNP, Extra MP=Lab (2)
 6  Aldershot:                   FPTP MP=C, Extra MP=Lab (2)
 7  Aldridge-Brownhills:         FPTP MP=C
 8  Altrincham and Sale West:    FPTP MP=C, Extra MP=Lab (2)
 9  Alyn and Deeside:            FPTP MP=Lab, Extra MP=C (2)
10  Amber Valley:                FPTP MP=C, Extra MP=Lab (2)
11  Angus:                       FPTP MP=SNP, Extra MP=C (2)
12  Arfon:                       FPTP MP=Lab, Extra MP=C (3) **
13  Argyll and Bute:             FPTP MP=SNP, Extra MP=C (2)
14  Arundel and South Downs:     FPTP MP=C, Extra MP=LD (2)
15  Ashfield:                    FPTP MP=C, Extra MP=Other (2)
16  Ashford:                     FPTP MP=C, Extra MP=Lab (2)
17  Ashton-under-Lyne:           FPTP MP=Lab, Extra MP=C (2)
18  Aylesbury:                   FPTP MP=C, Extra MP=LD (3) **
19  Ayr, Carrick and Cumnock:    FPTP MP=SNP, Extra MP=C (2)
20  Banbury:                     FPTP MP=C, Extra MP=LD (3) **
21  Banff and Buchan:            FPTP MP=C
22  Barking:                     FPTP MP=Lab, Extra MP=C (2)
23  Barnsley Central:            FPTP MP=Lab, Extra MP=Brexit (2)
24  Barnsley East:               FPTP MP=Lab, Extra MP=Brexit (2)
25  Barrow and Furness:          FPTP MP=C, Extra MP=Lab (2)
26  Basildon and Billericay:     FPTP MP=C
27  Basingstoke:                 FPTP MP=C, Extra MP=Lab (2)
28  Bassetlaw:                   FPTP MP=C, Extra MP=Brexit (3) **
29  Bath:                        FPTP MP=LD, Extra MP=C (2)
30  Batley and Spen:             FPTP MP=Lab, Extra MP=Other (3) **
31  Battersea:                   FPTP MP=Lab, Extra MP=LD (3) **
32  Beaconsfield:                FPTP MP=C, Extra MP=Other (2)
33  Beckenham:                   FPTP MP=C, Extra MP=LD (3) **
34  Bedford:                     FPTP MP=Lab, Extra MP=C (2)
```

Fig. 12.9 Selection of constituency MPs returned via First-Past-The-Post (FPTP) in the 2019 UK General Election. In addition are Extra MPs chosen in order to make up a Parliament that is (approximately) in proportion to national votes for each party. The number in brackets (...) is the position of the Extra MP in the contest. When the position is greater than second place, I have starred this result. These are the 'unintended consequences' that might prove both bizarre and unpopular to the electorate!

(2) Determine extra seats to make up PR proportions. My code for this is in the footnote.[17] The second and third line prevents **extra_seats** becoming negative. Note each named quantity is actually an array with rows corresponding to constituencies and columns corresponding to parties.

---

[17]%Calculate the seats that should have been won, if the number of seats is in proportion to the total national vote

```
pr_seats = round(total_vote_percent*sum(seats)/sum(total_vote_percent));
```
%Determine maximum ratio between seats and PR seats.
```
rmax = max(seats./pr_seats);
```
%Calculate extra seats
```
extra_seats = round(pr_seats*rmax - seats);
```
%Compute seat proportions and compare to PR ideal
```
new_seat_proportions = (extra_seats + seats)/sum(extra_seats + seats);
pr_proportions = pr_seats/sum(pr_seats);
```

(3) Remove the winning party from the constituency votes. Then for each party:

   (a) Sort the constituency votes in descending order, so we add extra seats to the constituencies where these parties gained the most votes.

   (b) Add the extra seats (i.e. keep track of additions in an array matching the dimensions above), one per constituency in order of the sorted list above, until all extra seats per party are allocated.

   (c) To prevent more than one extra MP per constituency, set the number of constituency votes for constituencies where extra MPs were added to be -1, so these appear at the bottom of the sorted list.

(4) Open a text file, and step through each constituency and print out something like: `Arfon: FPTP MP=Lab, Extra MP=C (3) **`. Print the position of the extra MP in brackets, and indicate with an extra `**` if the position was (3). These are the results which are likely to be less palatable.

(5) Save text file and close. Print overall statistics of seats, and compare to PR ideals.

12.3 Conclusion

This chapter is really all about transformation of *data* into *information*, and how *array*-based (i.e. spreadsheet) data structures can be manipulated in *code* to facilitate "what-if?" modelling. In this case we run a scenario of adding extra MPs (in a hopefully fair and reasonable way, i.e. subject to rules and constraints) to result in a more proportional representation of parties in the UK Parliament. Although my graphs are far cruder than the genuinely beautiful infographics that appear from broadcasters and journalists during elections, I hope they provide a balanced viewpoint, and in their variety, highlight the possibility of bias and misconceptions that certain representations (particularly those based upon percentages) may inadvertently suggest. I am satisfied that the practical details of a PR system, without violating the current precedent of local accountability of MPs to their constituency, *could in principle* be achieved, although I think a near doubling or Parliament, and a large number of third-place MPs would make this very hard to sell to the electorate. The nearly four years of national palsy following the EU vote in 2016 probably means another referendum of any sort is likely to be very unpopular anyway, and I can't imagine a scenario where a current government would actively promote a system that results in them having less parliamentary seats. In other words, I predict we will have a FPTP *status quo* for the foreseeable future, unless a major constitutional crisis or 'super hung parliament' (which requires a more complex coalition than in 2010) forces a more fundamental rethink of UK political structures. Attempting this analysis has possibly changed my own opinion that the existing system might not be as bad as I once thought. This is perhaps a valuable lesson in itself.

Appendix A

Linear Regression – Determining Lines of Best Fit

Perhaps the most important analytical tool in the physical sciences is the ability to quantify the validity of a mathematical model that relates a set of measurable parameters. The idea is as follows:

(1) Rearrange the model in such a way that it becomes a *linear equation* of the form $y = mx + c$.
(2) Plot experimental (x, y) data on a graph and determine the *line of best fit* through the data, using the $y = mx + c$ recipe described in the next section.
(3) Determine the gradient m and y intercept c, from the line of best fit.
(4) Alternatively, if you have good reason to believe a direct proportion between y and x, assume $y = mx$, and find m from a line of best fit using the $y = mx$ recipe.
(5) Determine the *uncertainty* in both m and c (or just m if performing a $y = mx$ fit).
(6) Determine a quantitative measure, r, of how good the fit is. This is called the *product moment correlation coefficient*. If $|r|$ is close to unity, then this means a strong correlation of model to measurement.

A.1 Fitting $y = mx + c$

The idea is to minimise the sum S of squared deviations of $\{x_i, y_i\}$ data from a line of best fit $y = mx + c$.

$$S = \sum_{i=1}^{N} (y_i - mx_i - c)^2 \qquad (A.1)$$

The minimum of S can be found by *partial* differentiation with respect to m and c, and setting these expressions to zero. We must use partial derivatives since S is a function of two variables, m and c. In a visual sense, imagine plotting a surface of height S above a Cartesian grid of m,c values, given a particular $\{x_i, y_i\}$ dataset, as in Fig. A.1. The optimal m,c pairing corresponds to the lowest point of this surface, i.e. where $\frac{\partial S}{\partial m} = 0$ and $\frac{\partial S}{\partial c} = 0$.

$$\frac{\partial S}{\partial m} = 2 \sum_{i=1}^{N} (y_i - mx_i - c)(-x_i) \tag{A.2}$$

$\frac{\partial S}{\partial m} = 0$ implies:

$$\sum_{i=1}^{N} \left(x_i y_i - mx_i^2 - cx_i\right) = 0$$

$$\frac{1}{N} \sum_{i=1}^{N} x_i y_i - m\frac{1}{N} \sum_{i=1}^{N} x_i^2 - c\frac{1}{N} \sum_{i=1}^{N} x_i = 0 \tag{A.3}$$

Define the following statistical averages associated with the $\{x_i, y_i\}$ data:

$$\bar{x} = \frac{1}{N} \sum_{i=1}^{N} x_i, \quad \bar{y} = \frac{1}{N} \sum_{i=1}^{N} y_i,$$

$$\overline{x^2} = \frac{1}{N} \sum_{i=1}^{N} x_i^2, \quad \overline{y^2} = \frac{1}{N} \sum_{i=1}^{N} y_i^2, \quad \overline{xy} = \frac{1}{N} \sum_{i=1}^{N} x_i y_i \tag{A.4}$$

Hence:

$$\overline{xy} - m\,\overline{x^2} - c\bar{x} = 0 \tag{A.5}$$

Now:

$$\frac{\partial S}{\partial c} = 2 \sum_{i=1}^{N} (y_i - mx_i - c)(-1) \tag{A.6}$$

therefore $\frac{\partial S}{\partial c} = 0$ implies:

$$\sum_{i=1}^{N} (y_i - mx_i - c) = 0$$

$$\sum_{i=1}^{N} y_i - m \sum_{i=1}^{N} x_i - Nc = 0$$

$$\bar{y} - m\bar{x} - c = 0 \tag{A.7}$$

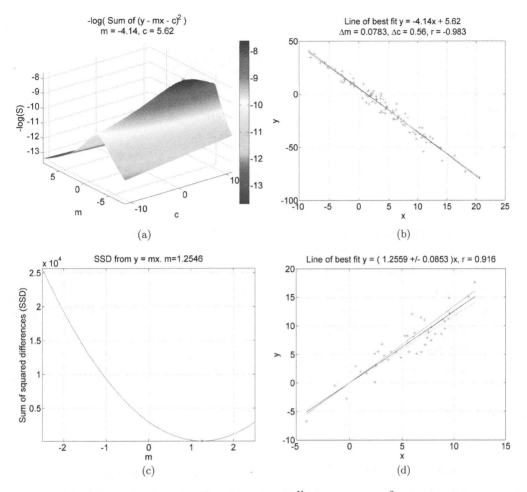

Fig. A.1 (a) Surface of $-\log_{10} S$, where $S(m, c) = \sum_{i=1}^{N} (y_i - mx_i - c)^2$, plotted against m and c values of potential lines of best fit $y = mx + c$ to a set of $\{x_i, y_i\}$ data. The peak of the $-\log_{10} S$ surface is the (m, c) pairing that minimises S and hence corresponds to the line of best fit. In this case $m = -4.14$ and $c = 5.62$. (b) Line of best fit $y = (-4.14 \pm 0.08) x + (5.62 \pm 0.56)$ underlays the $\{x_i, y_i\}$ data. The product moment correlation coefficient $r = -0.983$. (c) Plot of $S(m) = \sum_{i=1}^{N} (y_i - mx_i)^2$ vs m using a different $\{x_i, y_i\}$ dataset, that is assumed to include (0,0). The minimum S at $m = 1.26$ corresponds to the optimal line of of best fit $y = (1.26 \pm 0.09) x$ illustrated in (d). Note the $m = 1.2546$ value picked in (c) is simply the lowest $S(m)$ of those chosen, given a linearly spaced set between -3 and 3. The true minimum $\frac{\partial S}{\partial m} = 0$ is when $m = 1.2559$. The product moment correlation coefficient $r = 0.916$.

i.e.

$$\bar{y} = m\bar{x} + c \tag{A.8}$$

The result $\bar{y} = m\bar{x} + c$ implies a line of best fit $y = mx + c$ must pass through the mean average coordinates $(\bar{x}, \bar{y})$. Substituting for $c = \bar{y} - m\bar{x}$ into $\overline{xy} - m\,\overline{x^2} - c\bar{x} = 0$ yields an expression for m in terms of the statistical measures of our $\{x_i, y_i\}$ data set:

$$m = \frac{\overline{xy} - \bar{x}\bar{y}}{\overline{x^2} - \bar{x}^2} \tag{A.9}$$

Note if we define the *variance* $V[x] = \overline{x^2} - \bar{x}^2$ and *covariance* $\text{cov}[x, y] = \overline{xy} - \bar{x}\bar{y}$, we arrive at a slightly easier-to-remember expression:

$$m = \frac{\text{cov}[x, y]}{V[x]} \tag{A.10}$$

If we repeat the analysis for the line $x = My + d$, but consider squared deviations of x (rather than y) this time, then $M = \frac{\text{cov}[x,y]}{V[y]}$. If there is a perfect correlation between x and y then we would expect $x = My + d$ and $y = mx + c$ to be the *same line*, i.e. $mx + c = (1/M)(x - d)$. If this is true for all x then $mM = 1$. In reality this will not be the case, but the product $r^2 = mM = \frac{(\text{cov}[x,y])^2}{V[x]V[y]}$ will nonetheless yield a useful metric of correlation. A value of unity means a perfect fit, whereas a value closer to zero means no discernable correlation between x and y. Motivated by this idea, let us define the *product moment correlation coefficient*:

$$r = \frac{\text{cov}[x, y]}{\sqrt{V[x]V[y]}} \tag{A.11}$$

Note r is more useful than r^2 (i.e. the R^2 output of the Microsoft Excel trendline function) as the sign shows whether the line of best fit has a positive ($r > 0$) or negative ($r < 0$) gradient. The range of r is: $-1 \le r \le 1$.

It is possible to show[1]:

$$\Delta m = \frac{s}{\sqrt{N}} \frac{1}{\sqrt{V[x]}}, \quad \Delta c = \frac{s}{\sqrt{N}} \sqrt{1 + \frac{\bar{x}^2}{V[x]}} \tag{A.12}$$

where:

$$s = \sqrt{\frac{1}{N-2} \sum_{i=1}^{N} (y_i - mx_i - c)^2} \tag{A.13}$$

---

[1] https://mathworld.wolfram.com/LeastSquaresFitting.html.

s is the *unbiased estimator* of the standard deviation in the y values from the line of best fit. The $N - 2$ factor is due to *two* parameters (m and c) being used in the calculation, which are of course derived from the sample data themselves as shown above. Δm and Δc are very useful quantities, as they represent the *uncertainties* in m and c, and can therefore be used to calculate uncertainties in model parameters that are algebraically related to m and c.

So in summary, for a $y = mx + c$ line of best fit:

$$\bar{x} = \frac{1}{N} \sum_{i=1}^{N} x_i, \quad \bar{y} = \frac{1}{N} \sum_{i=1}^{N} y_i,$$

$$\overline{x^2} = \frac{1}{N} \sum_{i=1}^{N} x_i^2, \quad \overline{y^2} = \frac{1}{N} \sum_{i=1}^{N} y_i^2, \quad \overline{xy} = \frac{1}{N} \sum_{i=1}^{N} x_i y_i$$

$$V[x] = \overline{x^2} - \bar{x}^2, \quad V[y] = \overline{y^2} - \bar{y}^2, \quad \text{cov}[x, y] = \overline{xy} - \bar{x}\bar{y},$$

$$m = \frac{\text{cov}[x, y]}{V[x]}, \quad c = \bar{y} - m\bar{x}$$

$$r = \frac{\text{cov}[x, y]}{\sqrt{V[x]V[y]}}, \quad s = \sqrt{\frac{1}{N-2} \sum_{i=1}^{N} (y_i - mx_i - c)^2}$$

$$\Delta m = \frac{s}{\sqrt{N}} \frac{1}{\sqrt{V[x]}}, \quad \Delta c = \frac{s}{\sqrt{N}} \sqrt{1 + \frac{\bar{x}^2}{V[x]}} \tag{A.14}$$

A.2 Fitting $y = mx$

The idea is to minimise the sum S of squared deviations of $\{x_i, y_i\}$ data from a line of best fit $y = mx$, i.e. a *direct proportion* is asserted between x and y.

$$S = \sum_{i=1}^{N} (y_i - mx_i)^2 \tag{A.15}$$

$$\frac{\partial S}{\partial m} = 2 \sum_{i=1}^{N} (y_i - mx_i)(-x_i) \tag{A.16}$$

Therefore, $\frac{\partial S}{\partial m} = 0$ implies:

$$\frac{1}{N} \sum_{i=1}^{N} x_i y_i - m\frac{1}{N} \sum_{i=1}^{N} x_i^2 = 0 \tag{A.17}$$

i.e.

$$m = \frac{\overline{xy}}{\overline{x^2}} \tag{A.18}$$

The product moment correlation coefficient $r = \frac{\text{cov}[x,y]}{\sqrt{V[x]V[y]}}$ is the same as for the $y = mx+c$ fit, but the uncertainty in m is slightly different since only one parameter is used (i.e. m) in the computation of s. So in summary, for a $y = mx$ line of best fit:

$$\bar{x} = \frac{1}{N} \sum_{i=1}^{N} x_i, \quad \bar{y} = \frac{1}{N} \sum_{i=1}^{N} y_i,$$

$$\overline{x^2} = \frac{1}{N} \sum_{i=1}^{N} x_i^2, \quad \overline{y^2} = \frac{1}{N} \sum_{i=1}^{N} y_i^2, \quad \overline{xy} = \frac{1}{N} \sum_{i=1}^{N} x_i y_i$$

$$V[x] = \overline{x^2} - \bar{x}^2, \quad V[y] = \overline{y^2} - \bar{y}^2, \quad \text{cov}[x,y] = \overline{xy} - \bar{x}\bar{y},$$

$$m = \frac{\overline{xy}}{\overline{x^2}}$$

$$r = \frac{\text{cov}[x,y]}{\sqrt{V[x]V[y]}}$$

$$s = \sqrt{\frac{1}{N-1} \sum_{i=1}^{N} (y_i - mx_i)^2}$$

$$\Delta m = \frac{s}{\sqrt{N}} \frac{1}{\sqrt{V[x]}} \tag{A.19}$$

Appendix B

Using the Runge–Kutta Method to Solve the Pendulum Problem

B.1 The Runge–Kutta Method of Solving Ordinary Differential Equations

The *Runge–Kutta* finite step-size iterative numerical method for solving equations such as $\frac{dy}{dx} = f(x)$ trades more calculation steps for a tangible reduction in error. Whereas the *Euler Method* $y_{i+1} = y_i + f(x_i)\Delta x$ has errors that scale with Δx, Runge–Kutta[1] has errors that scale with Δx^4. So if you reduce the finite step size by a factor two, the Euler method error reduces by two, but Runge–Kutta reduces by a factor of sixteen. The recipe is as follows:

$$k_1 = f\left(x_i\right)$$

$$k_2 = f\left(x_i + \tfrac{1}{2}k_1\Delta x\right)$$

$$k_3 = f\left(x_i + \tfrac{1}{2}k_2\Delta x\right)$$

$$k_4 = f\left(x_i + k_3\Delta x\right)$$

$$x_{i+1} = x_i + \Delta x$$

$$y_{i+1} = y_i + \tfrac{1}{6}\Delta x\left(k_1 + 2k_2 + 2k_3 + k_4\right) \tag{B.1}$$

As with any differential equation solving scheme, the initial condition (x_0, y_0) needs to be defined, in addition to Δx and the number of iterations (or maximum value of x). The Runge–Kutta scheme is evaluated for the simple pendulum + drag, and the double-pendulums referred to in Chapter 5.

---

[1] *Runge–Kutta* methods are actually a *family* of recipes that can reduce errors even further, but with consequently even more calculation steps that can slow down the running of the algorithm in a computer simulation. In this book, I shall refer to the popular 'fourth order' Runge–Kutta variant.

B.2 Single Pendulum with Air Resistance

Equation for angular acceleration:

$$\ddot{\theta}(\theta, \dot{\theta}) = -4\pi^2 \sin\theta - \frac{\frac{1}{2}c_D \rho \pi r^2 l}{m} \dot{\theta} \left| \dot{\theta} \right|$$

Step 1:

$$A_1 = \dot{\theta}_n, \quad B_1 = \ddot{\theta}(\theta_n, \dot{\theta}_n), \quad \alpha = \theta_n + \tfrac{1}{2}A_1\Delta t, \quad w = \dot{\theta}_n + \tfrac{1}{2}B_1\Delta t \qquad \text{(B.2)}$$

Step 2:

Note intermediate parameters alpha and omega get redefined from step to step.

$$A_2 = w, \quad B_2 = \ddot{\theta}(\alpha, w), \quad \alpha = \theta_n + \tfrac{1}{2}A_2\Delta t, \quad w = \dot{\theta}_n + \tfrac{1}{2}B_2\Delta t \qquad \text{(B.3)}$$

Step 3:

$$A_3 = w, \quad B_3 = \ddot{\theta}(\alpha, w), \quad \alpha = \theta_n + A_3\Delta t, \quad w = \dot{\theta}_n + B_3\Delta t \qquad \text{(B.4)}$$

Step 4:

$$A_4 = w, \quad B_4 = \ddot{\theta}(\alpha, w) \qquad \text{(B.5)}$$

Update angles and angular velocities:

$$\theta_{n+1} = \theta_n + \tfrac{1}{6}\Delta t \left(A_1 + 2A_2 + 2A_3 + A_4\right) \qquad \text{(B.6)}$$

$$\dot{\theta}_{n+1} = \dot{\theta}_n + \tfrac{1}{6}\Delta t \left(B_1 + 2B_2 + 2B_3 + B_4\right) \qquad \text{(B.7)}$$

B.3 Double Pendulum without Air Resistance

Equation for angular acceleration:

$$\Delta = \phi - \theta$$

$$\ddot{\theta}(\theta, \phi, \dot{\theta}, \dot{\phi}) = \frac{m_2 l_1 \dot{\theta}^2 \sin\Delta \cos\Delta + m_2 g \sin\phi \cos\Delta + m_2 l_2 \dot{\phi}^2 - (m_1 + m_2)g\sin\theta}{(m_1 + m_2)l_1 - m_2 l_1 \cos^2\Delta}$$

$$\ddot{\phi}(\theta, \phi, \dot{\theta}, \dot{\phi}) = \frac{-m_2 l_2 \dot{\phi}^2 \sin\Delta \cos\Delta + (m_1 + m_2)\left(g\sin\theta\cos\Delta - l_1\dot{\theta}^2\sin\Delta - g\sin\phi\right)}{(m_1 + m_2)l_2 - m_2 l_2 \cos^2\Delta}$$

$$\text{(B.8)}$$

Step 1:

$$A_1 = \dot{\theta}_n, \quad B_1 = \dot{\phi}_n, \quad C_1 = \ddot{\theta}(\theta_n, \phi_n, \dot{\theta}_n, \dot{\phi}_n), \quad D_1 = \ddot{\phi}(\theta_n, \phi_n, \dot{\theta}_n, \dot{\phi}_n)$$

$$\alpha = \theta_n + \tfrac{1}{2}A_1\Delta t, \quad \beta = \phi_n + \tfrac{1}{2}B_1\Delta t, \quad \omega = \dot{\theta}_n + \tfrac{1}{2}C_1\Delta t, \quad \Omega = \dot{\phi}_n + \tfrac{1}{2}D_1\Delta t$$

Step 2:

$$A_2 = \omega, \quad B_2 = \Omega, \quad C_2 = \ddot{\theta}(\alpha, \beta, \omega, \Omega), \quad D_2 = \ddot{\phi}(\alpha, \beta, \omega, \Omega)$$

$$\alpha = \theta_n + \tfrac{1}{2}A_2\Delta t, \quad \beta = \phi_n + \tfrac{1}{2}B_2\Delta t, \quad \omega = \dot{\theta}_n + \tfrac{1}{2}C_2\Delta t, \quad \Omega = \dot{\phi}_n + \tfrac{1}{2}D_2\Delta t$$

Step 3:

$$A_3 = \omega, \quad B_3 = \Omega, \quad C_3 = \ddot{\theta}(\alpha, \beta, \omega, \Omega), \quad D_3 = \ddot{\phi}(\alpha, \beta, \omega, \Omega)$$

$$\alpha = \theta_n + A_3\Delta t, \quad \beta = \phi_n + B_3\Delta t, \quad \omega = \dot{\theta}_n + C_3\Delta t, \quad \Omega = \dot{\phi}_n + D_3\Delta t$$

Step 4:

$$A_4 = \omega, \quad B_4 = \Omega, \quad C_4 = \ddot{\theta}(\alpha, \beta, \omega, \Omega), \quad D_4 = \ddot{\phi}(\alpha, \beta, \omega, \Omega) \tag{B.9}$$

Update angles and angular velocities:

$$\theta_{n+1} = \theta_n + \tfrac{1}{6}\Delta t \left(A_1 + 2A_2 + 2A_3 + A_4\right) \tag{B.10}$$

$$\phi_{n+1} = \phi_n + \tfrac{1}{6}\Delta t \left(B_1 + 2B_2 + 2B_3 + B_4\right) \tag{B.11}$$

$$\dot{\theta}_{n+1} = \dot{\theta}_n + \tfrac{1}{6}\Delta t \left(C_1 + 2C_2 + 2C_3 + C_4\right) \tag{B.12}$$

$$\dot{\phi}_{n+1} = \dot{\phi}_n + \tfrac{1}{6}\Delta t \left(D_1 + 2D_2 + 2D_3 + D_4\right) \tag{B.13}$$

Afterword

Like any recipe book, *Science by Simulation* is best read actively. This means use the mathematical recipes as inspiration to construct spreadsheets and write computer code yourself. It should be well-thumbed and contain numerous annotations, chocolate sauce and coffee stains, although perhaps not if your copy has been borrowed from a library. If you feel adding a gastronomic marker is necessary for reasons of personalization and comfort, try not to obscure the mathematics. World Scientific will release an electronic version of this book, but if you wish to peruse the majority of MATLAB code and Excel sheets referred to in this publication, do please download the files from my website: http://www.eclecticon.info/scibysim. *The Eclecticon* also contains a fairly comprehensive collection of notes and resources that cover the majority of pre-University Physics and Mathematics, and a wide range of scientific computing examples. And a few other more arty things, too.

Bibliography

[1] Abbott *et al.* (2016) "Observation of gravitational waves from a binary black hole merger." *Phys. Rev. Lett.* **116**, 061102.

[2] Barry, R.G., Chorley, R.J. (2010) *Atmosphere, Weather and Climate*, 9th Edition, Routledge, pp. 30–37.

[3] Barton, J.T. (2016) *Models for Life: An Introduction to discrete Mathematical Modelling with Microsoft Office Excel*, Wiley, pp. 192–194.

[4] Brauer, F. (2004) *Mathematical Models in Population Biology And Epidemiology*, Springer.

[5] Cullerne, J.P., French, A., Poon, D., Baxter, A., Thompson, R.N. (2020) "The pedagogical power of context: Extending the epidemiology of Eyam." *Phys. Educ.* **55**.

[6] Dean, K.R., Krauer, F. *et al.* (2018) "Human ectoparasites and the spread of plague in Europe during the Second Pandemic." *PNAS* **115**(6), 1304–1309.

[7] Faber, T.E. (1997) *Fluid Dynamics for Physicists*, Cambridge University Press, p. 33.

[8] Ferguson, N.M. *et al.* (2020) "Impact of non-pharmaceutical interventions (NPIs) to reduce COVID-19 mortality and healthcare demand." Imperial College COVID-19 Response Team. 16 March 2020. https://doi.org/10.25561/77482.

[9] Feynman, R.P. (1999) *The Pleasure of Finding Things Out.* Penguin Books.

[10] Foster, R.E.J. (2019) *50 Treasures from Winchester College*, Scala.

[11] French, A., Kanchanasakdichai, O., Cullerne, J.P. (2019) "The pedagogical power of context: Iterative calculus methods and the epidemiology of Eyam." *Phys. Educ.* **54**, 045008.

[12] French, A., Cullerne, J.P., Kanchanasakdichai, O. (2019) "Numerical methods as an introduction to calculus." *Phys. Educ.* **54**, 045009.

[13] French, A., www.eclecticon.info.

[14] French, A. (2010) *Target Recognition Techniques for Multifunction Phased Array Radar*, Doctoral Thesis, Department of Electronic and Electrical Engineering. University College London.

[15] Gani, J. (1967) "On the general stochastic epidemic." In *Proceedings of the Fifth Berkeley Symposium on Mathematical Statistics and Probability*, Vol. 4: *Biology and problems of Health*, pp. 271–279, University of California Press, Berkeley, California.

[16] Jidong, G.J., Brewster, K., Xue, M. (2005) "A comparison of the radar ray path equations and approximations for use in radar data assimilation." *J. Atmos. Oceanic. Technol.* March 2005.

[17] Gleick, J. (1998) *Chaos*, Vintage.

[18] Gnädig, P., Honyek, G., Riley, K. (2001) *200 Puzzling Physics Problems*, Cambridge University Press.

[19] Hanselman, D., Littlefield, B. (2001) *Mastering MATLAB 6*, pp. 292–294. Prentice Hall, Inc.

[20] Hart, W., Hochfilzer, L., Cunniffe, N., Lee, H., Nishiura, H., Thompson, R.N. (2019) "Accurate forecasts of the effectiveness of interventions against Ebola may require models that account for variations in symptoms during infection." BioRxiv. 2019 Jan 1:592030.

[21] Hecht, E. (1998) *Optics*, 3rd Edition, Addison-Wesley.

[22] King, D.A., Rashed, R., *et al.* (2016) *Astronomy and Islamic society: Qibla, gnomics and timekeeping*, Encyclopedia of the History of Arabic Science. Volume 1. pp. 128–184. ISBN 978-0203711842. Retrieved 10 November 2016.

[23] Kermack, W.O., McKendrick, A.G. (1927) "A contribution to the mathematical theory of epidemics." *Proc. Roy. Soc. London* **115**(A), 700–721.

[24] Kendall, D.G. (1956) "Deterministic and stochastic epidemics in closed populations." In *Proceedings of the Third Berkeley Symposium on Mathematical Statistics and Probability, Volume 4: Contributions to Biology and Problems of Health*, pp. 149–165, University of California Press, Berkeley, California.

[25] Lewis, M. (2014) *Flashboys: A Wall Street Revolt*, W.W. Norton & Company.

[26] Lloyd, J. (2007) *A Concise Guide to Weather*, Paragon Books.

[27] Mackowiak, P.A., Sehdev, P.S. (2002) "The origin of quarantine." *Clin. Infect. Dis.* **35**(9), 1071–1072.

[28] Mandelbrot, B.B. (1982) *The Fractal Geometry of Nature*, Holt & Company.

[29] Morin, D. (2007) *Introduction to Classical Mechanics*. Cambridge University Press.

[30] Nakayama, Y., Boucher, R.F. (1999) *Introduction to Fluid Mechanics*, Butterworth-Heinemann.

[31] Oyston, P. (2001) "Plague virulence." *J. Med. Microbiol.* **50**, 1015–1017.

[32] Pandit, J.J. (2020) "Managing the R_0 of COVID-19: Mathematics fights back." *Anaesthesia* doi: 10.1111/anae.15151.

[33] Phillips, A.W. (1950) *Economica, New Series* **17**(67), 283–305.

[34] Press, W.H., Teukolsky, S.A., Vetterling, W.T., Flannery, B.P. (2003) *Numerical Recipes in C++. The Art of Scientific Computing*, 2nd Edition, Cambridge University Press.

[35] Rachah, A., Torres, D.F.M. (2015) "Mathematical Modelling, Simulation and Optimal Control of 2014 Ebola Outbreak in West Africa." *Discrete Dynam. Nat. Soc.* 2015, Article ID 842797. Hindawi Publishing Corporation.

[36] Rasmussen *et al.* (2015) "Early divergent strains of yersinia pestis in eurasia 5,000 years ago." *Cell.* **163**, 571–582. http://dx.doi.org/10.1016/j.cell.2015.10.009

[37] Raworth, K. (2017) *Doughnut Economics: Seven Ways to Think Like a 21st-Century Economist*, Penguin Random House. ISBN 9781847941398, pp. 63–66.

[38] Rees, W.G. (1994) *Physics by Example*. Cambridge University Press.

[39] Richards, M.A. (2005) *Fundamentals of Radar Signal Processing*, McGraw-Hill Electronic Engineering.

[40] Rogers, S. (2013) *Facts are Sacred*, Guardian Books/Faber & Faber.

[41] Max, R., Hannah, R., Esteban, O.-O., Joe, H. (2020) "Coronavirus pandemic (COVID-19)." Published online at https://ourworldindata.org/coronavirus. Data downloaded on 5-Aug-2020.

[42] Rosling, H. (2018) *Factfulness*, Sceptre Books.

[43] Shaw *et al.* (1986) "Chaos." *Sci. Am.* **54**(12), 46–57.

[44] Silver, N. (2013) *The Signal and the Noise.* Penguin Books.

[45] Skiadas, C.H., Skiadas, C. (2009) *Chaotic Modelling and Simulation*, CRC Press.

[46] Skolnik, M.I. (2001) *Introduction to Radar Systems*, 3rd Edition, McGraw-Hill.

[47] Skolnik, M.I. (1990) *Radar Handbook.* Second Edition. McGraw-Hill.

[48] Stahl-Timmins for The BMJ. (2014) Infographic: Ebola: A clinical guide.

[49] Strogatz, S.H. (2000) *Nonlinear Dynamics and Chaos.* Westview Press.

[50] Sullivan, N. *et al.* (2007) "Ebola virus pathogenesis: Implications for vaccines and therapies." *J. Virol.* **77**(18), 9733–9737.

[51] Sutherland, S. (2007) *Irrationality*, 3rd Edition, Pinter & Martin.

[52] Tait, P. (2005) *Introduction to Radar Target Recognition*, The Institution of Electrical Engineers, IEE.

[53] Taylor, D. (2015) *The Roses of Eyam*, Samuel French. ISBN 0-573-11386-6.

[54] Thompson, R.N., Gilligan, C.A., Cunniffe, N.J. (2016) "Detecting pre-symptomatic infection is necessary to forecast major epidemics in the earliest stages of infectious disease outbreaks." *PLoS Comput. Biol.* **12**(4), e1004836.

[55] Thompson, R.N., Hart, W.S. (2018) "Effect of confusing symptoms and infectiousness on forecasting and control of Ebola outbreaks." *Clin. Infecti. Dis.* **67**(9), 1472–1474.

[56] Verity, R., *et al.* (2020) "Estimates of the severity of coronavirus disease 2019: A model-based analysis." *Lancet Infect. Dis.* **20**. 669–677. Published Online March 30, 2020.

[57] Wander, T. (2012) *Marconi on the Isle of Wight.* New Generation Publishing.

[58] Whittles, L.K., Didelot, X. (2016) "Epidemiological analysis of the Eyam plague outbreak of 1665–1666." *Proc. R. Soc. B.* **283**, 20160618. http://dx.doi.org/10.1098/rspb. 2016.0618, https://www.ncbi.nlm.nih.gov/pmc/articles/PMC4874723/.

[59] WHO Ebola Response Team. (2014) "Ebola Virus Disease in West Africa-The First 9 Months of the Epidemic and Forward Projections." *The New Engl. J. Med.* **371**(16).

[60] WHO Ebola Response Roadmap Situation Report, 29 October 2014.

[61] Wu, J.T. (2020) "Estimating clinical severity of COVID-19 from the transmission dynamics in Wuhan, China." *Nat. Med.* **26**, 506–510.

[62] Eyam and The Great Plague website. https://eyamandthegreatplague.weebly.com/ index.html (accessed 20/8/2018).

[63] The Plague: Types. Causes & Symptoms. https://www.healthline.com.health/ plague#symptoms (accessed 20/8/2018).

[64] Wikipedia: Edward Jenner. https://en.wikipedia.org/wiki/Edward_Jenner (accessed 1/9/2018).

[65] Wikipedia: Lady Mary Wortley Montagu https://en.wikipedia.org/wiki/Lady_Mary_ Wortley_Montagu (accessed 1/9/2018).

[66] Wikipedia: Great Plague of London. https://en.wikipedia.org/wiki/Great_Plague_of_ London (accessed 1/8/2018).

[67] BBC News. "Eyam Plague: The Village of the damned." http://www.bbc.co.uk/news/ uk-england-35064071 (accessed 20/8/2018).

[68] Wikipedia: Yersinia Pestis. https://en.wikipedia.org/wiki/Yersinia_pestis (accessed 1/8/2018).

Index

2D phased array, 172

CPSIA information can be obtained
at www.ICGtesting.com
Printed in the USA
JSHW051909110622
26964JS00002B/15

9 781800 611214